ATLAS RISING

AYN RAND AND SILICON VALLEY

ATLAS RISING AYN RAND AND SILICON VALLEY

Copyright © 2018 by the Atlas Rising Institute. All rights reserved.

ISBN 978-1-942868-026

The Atlas Rising Institute is a new educational organization dedicated to the study of creative human intelligence. Contact AtlasRising.org@gmail.com for more information.

Editorial note:

Rather than load this report with footnotes and references as would have been the norm at one time, Google and Wikipedia now take care of that chore for us. Just go online to find more information on the people, projects, achievements, problems, and sources cited throughout this document.

Thanks:

To David Kelley, founder of The Atlas Society, who understood that Ayn Rand's Objectivism was a work in progress and brought new levels of enlightenment to Objectivist scholarship.

To Jennifer Grossman, who is so capably carrying the torch as CEO of the Atlas Society.

To the Ayn Rand Institute for its dedicated work in assuring the widest possible distribution of Ayn Rand's writings.

To the thousands of admirers of Ayn Rand who work daily to apply the best of her philosophy in life and work. When they succeed, we're all the beneficiaries, as we see throughout this report.

To those who demonstrate the toleration, benevolence, and openness to new ideas that are essential to human progress. Those Objectivists who are best informed, acknowledge that they have, and will, make mistakes. But they know that freedom and the universal expansion of the power of the rational human mind are essential to our survival and the best possible future for all.

TABLE OF CONTENTS

INTRODUCTION   1
A Warning Note

CHAPTER ONE   3
Those Who "Have Read the Book"

CHAPTER TWO   8
Breaking Free

CHAPTER THREE   13
Disinhibiting the Mind

CHAPTER FOUR   17
Breaking Bad Laws

CHAPTER FIVE   21
Thinking About Thinking

CHAPTER SIX   24
Private Space Exploration, Investment, and Development

CHAPTER SEVEN   26
Life Extension

CHAPTER EIGHT   28
Liberating Education, Part One – the Early Years

CHAPTER NINE   31
Liberating Education, Part Two – Higher.Edu

CHAPTER TEN   36
Environmental Sustainability

CHAPTER ELEVEN   38
Overthrowing Dictators and Authoritarian Governments

CHAPTER TWELVE   43
Enhancing the Human Brain, Expanding Artificial Intelligence

CHAPTER THIRTEEN   50
"Real People?"

CHAPTER FOURTEEN   51
The Future Isn't What It Used To Be . . . It's More

CHAPTER FIFTEEN   53
Oh, Oh, The Dark Side

CHAPTER SIXTEEN   56
A Case Study of an Objectivist Mode of Thinking

CHAPTER SEVENTEEN   60
An Objectivist Interviews a Whistleblower

CHAPTER EIGHTEEN   68
Reasonable Expectations

APPENDIX   69
Resources

Published by The Atlas Rising Institute (www.atlasrising.org)

INTRODUCTION   A Warning Note

Can someone explain the vitriol whenever Ayn Rand comes up?
"Atlas" is the greatest motivator for the individual that I can imagine.
Rob Rowe, film producer,
Disney/Pixar

Articles about Silicon Valley techno-entrepreneurs often include antagonistic remarks about Ayn Rand along with suggestions that Rand and the entrepreneurs are a menace and need to be, as some have suggested, "reigned in."

Those who have read Ayn Rand find the criticisms to be baffling. She said nothing like what she is often accused of saying. Her arguments for ethical self-interest, for legal systems that protect human rights, and for a free society have been grossly twisted and misstated. Those who are most vociferous in their condemnations of Rand are the most grossly uninformed as to what she actually said.

Rand's primary message: To fulfill our potential as human beings means being free to utilize our uniquely human means of survival and self-advancement – the functioning mind. That requires education in how to use the tools of reason, logic, and scientific method. It requires protection of the right to think freely, to create, exchange ideas, trade services, choose one's collaborators, choose one's ambitions, and be free above all to pursue one's dreams.

Those rights and pursuits have been forbidden to most of humanity for most of human history. Forbidden by superstition, authoritarianism, and anti-reason doctrines. Great libraries have been torched, scholars murdered, and people enslaved and slaughtered due to their class, race, and beliefs. These are ongoing conditions and there are many who defend the superstition, authoritarianism, and anti-rationality behind it all.

Following the American Revolution, the U.S. and much of the world is still gradually approaching universal liberation. That means the liberation of the mind and its benefits. This is a process that is accelerating rapidly in the creative atmosphere of Silicon Valley and similar environs.

Rand's advocacy of rational self-interest is the most commonly misunderstood and misstated aspect of her work. Rand's ethics are a continuation of Western ethical tradition starting with ancient Greece – particularly Aristotle and the Stoics of Greece and Rome. Her critics find that tradition to be reprehensible in terms of conflict with their ideals of subordinating individual self-interest to the assumed interests of society, the state, or the dictates of a deity or "higher reality."

Ayn Rand's fiction depicts the conflict of collectivism versus individualism. The collectivists – left-wing, right-wing, or religious – don't come out looking so good. Any of her readers who see their worldview expressed as conniving villainy are unlikely to go thumbs up.

Her nonfiction writing in defense of reason, enlightened self-interest and capitalism is unmatched in terms of detailed historical context and evidence-based argument.

Those who resonate with her writings often experience an integrated worldview for the first time in their lives. People who apply focused systematic thinking rather than random stream of consciousness in their lives are inspired to find someone who actually thinks that thinking and creative achievement are virtues. It's especially a relief to thoughtful young people who find themselves surrounded and baffled by the behavior of irrational and erratic adults.

Thus the attraction and influence that has resulted in a worldwide Libertarian movement founded by her admirers. And the deep influence she has in today's technological/entrepreneurial revolution, the influence of the system of thinking she named Objectivism.

Be forewarned that if you agree with major points of this document, you will be considered by some to be a vile person. If you're a student you'll be dismissed as a mis-guided adolescent. If you agree with and implement, the ideas herein, you'll likely be liberated to be more creative and productive than you would be otherwise. And if successful, will be considered to be "lucky." There's more, as you will see.

CHAPTER ONE   Those Who "Have Read the Book"

What is worshipped in Atlas Shrugged isn't money, it's production. It's accomplishment.
Jeff Rouner, journalist

What is it in Ayn Rand's writings that inspires so many of today's innovators in enterprise, education, advanced science, and technology?

Brad Keywell, cofounder of Groupon on the 1957 novel: *Atlas Shrugged creates a new brand of hero — one that embodies dedication, creativity, capitalistic achievement and excellence. The book defines a person's greatness by his or her dogged pursuit of entrepreneurial success and refusal to accept political barriers, pessimism and the status quo.*

There is a power in creators — of revolutionary products, designs, medicine, architecture and more — but most of our society has a default setting to be skeptical of these talented people and doubt their motives. The book caused me to also understand that some philosophies can be inhibitors, . . .

PayPal cofounder Luke Nosk (also of Founders Fund and a director at Space X): *. . . the most important lesson I got from Rand was that business can be good or evil. I have realized more and more that great companies, founded for a long-term purpose, such as Google or Facebook or SpaceX, may do more good in the world than any other vehicle that we have.*

James Crawford, cofounder CEO, Fieldbook: *That deep, abiding optimism, that sense of the grandeur of life and the world, is part of what drew me to Objectivism and what resonates with me so deeply. I think it's what resonates with others in the startup community, too.*

Ayn Rand has influenced founders and high-level participants throughout today's leading techno enterprises. The list that follows is a partial sampling of the diverse people who have been attracted to her thinking.

Those impacted by Ayn Rand include:

• Jimmy Wales, cofounder, Wikipedia. His goal is to provide free access to all the world's knowledge. (More follows later on Wales' work and his Objectivist roots.)

• Hal Finney, pioneer of BitCoin (considered by many to be the originator of the code).

• Michael Stern Hart, inventor of the eBook.

• Jeffrey Skolls, cofounder of eBay.

• Elon Musk, founder of Tesla Motors and Space X, referred to as "having read the book" (*Atlas Shrugged*).

• Larry Ellison, founder of Oracle.

• T.J. Rogers, founder Cypress Semiconductor.

• Marc Emery, dedicated Objectivist and much persecuted ideological force behind the pot legalization movement.

• Jason Crawford, cofounder & CEO, Fieldbook, plus management roles at Groupon and Amazon. Creator of the website Free Objectivist Books for Students.

• Peter Diamandis, founder/cofounder, Singularity University, the X Prize Foundation, Human Longevity Inc., Planetary Enterprises, Celularity, and others. Peter was named one of the world's 50 greatest leaders by Fortune Magazine.

• David Ebersman, CFO, Facebook.

• Evan Williams, cofounder, Twitter and Blogger.

• John Dorsey, cofounder, Twitter.

• Kevin Systrom, cofounder, Instagram.

• Alexander McCobin, CEO of Conscious Capitalism. Founder of Students for Liberty – Objectivist/Libertarian international free-market student organization.

• Edward Snowden, National Security Agency whistleblower.

• Chelsea Manning, human rights advocate and U.S. government whistleblower.

• Stewart Brand, creator of the *Whole Earth Catalog*. (Published and collaborated with Ted Nelson and other internet pioneers. Credited by Steve Jobs as a source of inspiration.)

• Ray Dalio, Bridgewater Associates. (World's largest hedge fund.)

- Evan Spiegel, cofounder of Snapchat.

- Steve Perlman, founder QuickTime, WebTV, Rearden Limited and Artemes Networks.

- Ben Best, former Director of the Cryonics Institute, currently Director of Research with the Life Extension Foundation.

- Max More, CEO, Alcor, Transhumanist and life-extension pioneer.

- Steve Forbes, national financial news publisher.

- John Stossel, national TV commentator.

- Peter Thiel, cofounder of PayPal and Founders Fund. Thiel while not explicitly an Objectivist, has been a long time supporter of Libertarian causes. (The Libertarian Party while not accepted by Ayn Rand, was founded and led by dedicated Objectivists.)

- Jeff Bezos, Amazon. Also, while not explicitly a supporter of Ayn Rand, is a long-time supporter of *Reason Magazine*, the leading Libertarian news source founded by Robert Poole and other Objectivists.

- Robert Poole, a cofounder of *Reason Magazine* and a pioneer in the worldwide movement toward privatizing what have traditionally been government monopolies.

- Rob Rowe, film producer, Disney/Pixar

- Bill Cockayne, cofounder of Scout Electromedia.

- Robert Frasca, Galt Technologies.

- Chip Wilson, founder of Lululemon Athletica.

- Zoltan Istvan, writer for National Geographic, sci-fi novelist, Transhumanist, Libertarian political candidate.

- Nathaniel Branden, Ayn Rand's chief associate prior to a bitter falling out. Branden founded the self-esteem movement, which while often superficially imitated, has had enduring positive results in education and psychotherapy.

- Steve Jobs, never explicitly Objectivist but also believed by Steve Wozniak to have "read the book," and beyond that, lived the spirit of the unstoppable intellectual pioneer.

- Dr. John Hospers, Ayn Rand acolyte, head of the Philosophy Department at the University of Southern California (USC), cofounder of the Libertarian Party, and first Libertarian Party presidential candidate.

- Ishan Gupta, Managing Director, Udacity.

- Travis Kalanick, cofounder of Uber.

- John Mackey, founder of Whole Foods, recently sold to Amazon. Mackey recently founded Conscious Capitalism a free-market advocacy organization.

Not all who have been influenced by Ayn Rand wish to continue saying so.

Travis Kalanick, for example, was previously quoted in the *Washington Post*, saying that *The Fountainhead* was one of his one of is favorite books. He "also brought up *Atlas Shrugged*, suggesting that the regulatory hellscape conjured by Rand bore an 'uncanny resemblance' to what Uber faced." But he has since disavowed Rand's influence.

John Mackey, has also declared some major differences with Objectivism in public debates and prefers not to be identified with the movement.

More from Brad Keywell of Groupon on *Atlas Shrugged*: *Before Atlas Shrugged, I didn't have an appreciation of philosophy or the imperative of having an explicit moral framework in life. But the book directed me toward insights about making the most of our limited time on earth, such as Aristotle's conviction that the pursuit of happiness is at the core of human existence, and that the good life is one of personal fulfillment.*

Along with the many thoughtful, deeply felt endorsements, there's a cultural war that rages regarding Ayn Rand. It's one that has caused some who have declared their respect for her to withdraw any such public pronouncement. They get a lot of flack which is so off-the-wall in their accusations as to be impossible to respond to in any rational manner other than saying, "You're mistaken, that's not what she said." That leads nowhere and responding to troll attacks just intensifies the angry assaults.

If you Google "Ayn Rand" you'll find her described as an advocate of unbridled greed, hater of the poor, worshiper of the rich, etc. Thus, if you express appreciation for her work, you're guilty too.

Most who are attracted to her thought, are thoughtful people who pursue multiple values and causes such as women's professional equality, environmentalism, philanthropy, civil rights, legal reforms, free worldwide education, and other "non-selfish" causes. These people bring intense questioning of the status quo and attitudes of un-limited possibilities to their work. The results, as you'll see throughout this report, are rather astounding.

After millennia of repression, human intelligence remains only barely understood or under conscious control. You'll see what even partially liberated intelligence can do throughout this text. But we're still miles away from understanding and using the brain's full potential.

Estimates of brain power vary. One estimate is 38 thousand trillion operations per second. Researchers at Duke University estimate a "low" of 100 trillion synaptic operations per second. No matter what estimate is correct, the number of operations going on in your brain at any instant is staggering. With that kind of computing power in our skulls, all that's missing is understanding how to fully use it – clearly a work in progress.

CHAPTER TWO   Breaking Free

> *"When someone says something is impossible, I stop listening."*
> Jeffrey Skolls, cofounder of eBay.

A questioning, unlimited possibilities attitude is the spirit that motivated a teenage Russian girl who grew up in the early days of the Bolshevik takeover.

With a total lack of practical realism, she thought she could escape the Soviet Union, get to the U.S., move to Hollywood and write scripts for movies.

She did escape, barely, in 1926 just as the gates were closing. She joined relatives in Chicago. They could only roll their eyes in disbelief when she told them her ambitions.

She went to Hollywood, worked as a waitress, took menial jobs at MGM, wrote scripts, married a handsome aspiring actor and started writing novels – working the dream, as it were. She was also examining her own beliefs and the methods for judging her thinking and values. Thus did she also become an avid reader of philosophy.

She would go on to write novels about engineers, architects, inventors, artists, and entrepreneurs – all of whom would succeed against immense odds. And she wrote about critics, competitors, and politicians who wanted to discredit and block the achievers, or steal from them.

Along the way she found the roots of the world's problems, as she had personally experienced them in the Soviet revolution, in the works of the mystic-minded Neoplatonic philosophers such as Hegel, and later, Martin Heidegger, a scholarly defender of the Nazi Party. And August Comte who coined the term "altruism" to describe the only moral life as total self-sacrifice in "living for others."

Hegel gave the world *"Pure Being and pure Nothing are thus one and the same,"* and argued why the State is superior to the individual. And there were those of the German, British, and U.S. philosophic advocates of Idealism who argued whether reality or human consciousness actually existed. These are philosophers whose anti-reason idiologies continue to strongly influence Western culture.

In contrast, Rand found philosophic inspiration in Aristotle. Aristotle held that the world is as we perceive it, not a shadow of a supernatural reality. He invented systematic logical thinking, scientific method, and an ethic based on happiness as the proper goal of human existence – happiness being the result of dedication to learning, virtue, and achievement of one's values. His work has been much misstated in translations by advocates of Catholic theology as well as by philosophic adversaries. Aristotle has long been denounced by those who believe in philosophic mysticism. By advocating reason-based Aristotelianism, Rand set herself up for the same kind of opposition and denigration.

Millions have read Ayn Rand's novels. Fewer have read her essays on philosophy, psychology, and economics. Considerably fewer still have read her most important philosophical treatises. (See the Appendix for sources.)

So there are variations and levels of understanding of her thought. For many, Rand and the philosophy she called Objectivism, is just one of many influences. Others experienced a life turnaround due to her work and have gone on to advocate and elaborate on her thinking.

Some people read Ayn Rand's novels and interpret their message to match their preexisting belief systems. So some can say they're influenced or inspired by Rand, but the benefits are expressed in limited and sometimes negative ways.

Rand published many essays to elaborate on the ideas behind her novels. But her most important and useful works dealt with Objectivist ethics and epistemology – the philosophy of mind as derived from Aristotle.

Her works were especially welcomed by people who were baffled and suffered in their youth in a world of hypocrites, bigots, and advocates of irrational belief systems.

Many of our brightest children are emotionally battered by parents, relatives, teachers, peers, and religious advocates who behave irrationally and excuse their daily self-contradictions as an uncontrollable part of the human condition.

The world of popular media doesn't help. TV and movie characterizations of creative geniuses usually depict them as suffering personality disorders or outright mental illness. People who are shown having ambitious goals and making plans are generally up to no good. Business people are shown as takers of other people's money and who spend most of their time "making deals" rather than producing anything worthwhile. "Normal" good people – the main characters who inhabit novels and movies – are usually depicted as screaming and swearing at their problems instead of thinking about them and solving them.

So imagine the dilemma of young thinkers who are alone in trying to figure out how to survive in an irrational world. "Is That All There Is?" as the bitter Peggy Lee song lyric asked.

Then, a bright spot, "the book." A book comes along about people who think about what they're doing and don't contradict their values. Imagine, heroes in novels who have a clear understanding of problems in the world around them and work to fix what's wrong. (They don't, incidentally, solve problems in the dominant comic book and video game fashion, by beating them up or shooting them.)

Many who admire Rand start with mixed attitudes that include a pro-reason, pro-creativity mindset, a mindset much disdained by most people around them. They find emotional and intellectual support in Rand's novels for their ways of thinking and their dreams. If they also read Rand's nonfiction essays, they learn the historical origins of contrary ideologies, those that uphold emotion and superstition as superior to reason.

After reading Rand, when young people question how things are done in the world around them, they don't have to take seriously the dark advice of others who will assure them that "It's just the way things are and no way are you going to change it." They can listen to their inner voice that says: "You can do this. You can fix the world. You have the ability. You can figure things out and do it."

This spark is expressed as a recurring theme among the Valley's thinker/doers such as Jeffrey Skoll: "When someone says something is impossible, I stop listening."

Peter Thiel, comments on those who oppose even considering doing impossible things: "We don't need to really worry about those people very much, because since they don't think it's possible they won't take us very seriously. And they will not actually try to stop us until it's too late."

Consider this remarkably insightful statement by Elon Musk who, according to his "assistant number 14" who lectured for an Objectivist audience, has read Atlas Shrugged.

Musk's ultimate goal is to ". . . expand the scope and scale of consciousness and knowledge." A worthy ambition and he zeros in on the basics of problem-solving as follows:

"To the degree that we can better understand the universe, then we can better know what questions to ask. Then whatever the question is that most approximates: what's the meaning of life?

"That's the question we can ultimately get closer to understanding. And so I thought to the degree that we can expand the scope and scale of consciousness and knowledge, then that would be a good thing."

This is the start of uninhibited thinking at its finest.

That's what Ayn Rand did when thinking out the basic principles of Objectivism.

And that's what Objectivism has done for many, giving them the spark to break free – get to fundamentals and start questioning their values, assumptions and mental processes that underlie everything they think and do.

They also achieve a new perspective of their "**self**" as being their **mind**. "Selfishness" in Aristotelian and Objectivist terms is means pursuing the benefits of consistent application of one's rational and creative mental faculties.

Those not acquainted with Neoplatonic and other subjectivist, anti-Aristotelian philosophies may not realize how deeply felt and influential those beliefs are.

For example, a classic textbook published years back on the *Philosophy of Education*, by a Yale professor dealt with the dilemma of teaching students about dealing with life and the world when existence itself is in doubt. He asks how can a teacher explain to a child who lives in a home, goes to school, and walks along streets. . . . how explain that there's no way of proving their existence?

Educators, according to this treatise, have to deal with the world of nature that "is but a defection from a changeless supernatural world, . . . " In fairness, the author isn't necessarily advocating anti-existence philosophies, he's just stating the problem it poses for educators.

It's an ancient idea that leads to inevitable contradictions. If you don't exist, for example, why bother eating non-existing food. A friend of Ayn Rand was once described as talking to a student who questioned his own existence: "I hope you figure it out soon," she said, "otherwise I'll be standing here talking to thin air."

If existence, or the capability of the mind to grasp "real" existence, is in doubt, then one must rely on the judgments of those who claim to be in touch with the higher reality – become a believer instead of a thinker. This is mindset of those who surrender their judgment (their minds) to religion, cults, collectivism, and dictatorship – all as well detailed in the book *The True Believer,* by Eric Hoffer.

CHAPTER THREE   Disinhibiting the Mind

We need to challenge the presumptions that whatever we're doing right now, the status quo, is the best of all possible worlds.

Edward Snowden

A major source of talent for Silicon Valley is the Massachusetts Institute of Technology. MIT is also a source of many of today's liveliest technologies such as Artificial Intelligence, Virtual Reality, Space Sciences, and Nanotechnology. So it's no great surprise to find the student bookstore racks stacked end-to-end with Ayn Rand's books – intellectual treats and motivators for ambitious young thinkers and doers who found very little support or understanding from others – family, teachers or peers – during childhood.

An example of MIT's intellectual resources is the late Dr. Amar G. Bose, founder of the Bose Corporation.

He said: "If I tell you that 'better' inspires fear – that even in the corporate world, people are scared of something better, you'd say that's ridiculous; everybody wants something better. Well, something better is always different. It isn't possible to make something better that isn't different. Whatever it is, if it's exactly the same, it isn't better. So it's the 'different' that scares people. When something's different, it's a heck of a gamble. And that's where 'courage' comes in."

That's the essence behind Steve Job's Apple motto: "Think different!" When you "think different," you'll be on notice your efforts will not necessarily be well received.

For example, when Ted Nelson, creator of hypertext, conceived what could have become a variant on today's World Wide Web or Wikipedia, he first received financing for his project (called "Xanadu") from a founder of AutoDesk. Later when a new CEO of AutoDesk came across "Xanadu" in the budget, she saw it as weird and expendable.

Similarly, Tim Berners-Lee's first efforts to create a cross-platform to integrate divergent computer operating systems received no encouragement.

His peers commented: "Who needs it?" "Impossible waste of time." "Pointless." Berners Lee used computer resources at CERN in Switzerland (under somewhat false pretenses, since the higher-ups weren't likely to approve such a pointless project) and created the World Wide Web which, strangely enough, now has no doubters whatsoever.

K. Eric Drexler and Christine Peterson left MIT for Silicon Valley in 1986 to promulgate education and research in nanotechnology. Objectivists in Berkeley were among the first to review Eric's draft manuscript for *Engines of Creation*, help promote student study groups and publish the first newsletters of Drexler and Peterson's newly founded Foresight Institute. The Foresight Institute strategy was to educate graduate students who would go on to teach and eventually head up university departments. It worked to the point where every major university in the world now has a nanotechnology initiative. Unfortunately much current nanotech research has gone off track.

The original concept of Nanotechnology, molecular manufacturing, became sabotaged by opponents. Funding for research in the original essence of the concept – using atoms as building blocks, to build molecules to make anything you can imagine – has declined considerably. Fortunately, some of our brightest minds are still working to bring the original vision into reality. Thanks to them and the Foresight Institute, it's just a matter of time.

Time and time again, a radical idea takes hold and then becomes hijacked or derailed. Or a great idea may become a corporate success and then become administered by the same kind of Board of Directors who would have rejected the idea in the first place.

When "normal" people in corporations are introduced to a new concept that replaces the old ones, the resistance is automatic and inevitable.

For example, the inventors of the photocopy machine first pitched their invention to IBM. IBM conducted market research that showed expensive, noisy copier machines couldn't compete with low-cost, convenient typewriter carbon paper. So no deal.

So Xerox went on its own; and much later had the luxury of creating a think tank, Xerox Parc, where people could invent wonderful things. And they did.

Guest innovators at Xerox Parc invented the graphic interface, computer mouse, and other components that made the original Apple computer possible.

But executives at Xerox, like IBM and other giant computer companies of the time, scoffed at the idea of a computer the size of a typewriter that could compete with the multi-million dollar behemoth computers that were top of the line in those days.

Desktop terminals linked to giant mainframes could handle any corporate chore one could throw at them, so the idea of making small computers was a pointless waste of time. It was widely believed that six mainframe computers would be sufficient to handle all the world's computing needs.

The personal computer wasn't taken seriously enough to worry about. It was left to you-know-who and friends at Apple who proceeded to change the world and drive some of the giants of the industry out of business.

From an interview with Apple cofounder Steve Wozniak: "Despite times when Apple was in financial and structural turmoil, Wozniak believes Jobs' speed of thought and endless drive helped the company move forward, believing that he may have adopted the ethos of the hard-working, never-failing Hank Reardon in Ayn Rand's book *Atlas Shrugged*. 'Steve was very fast thinking and wanted to do things, I wanted to build things. I think *Atlas Shrugged* was one of his guides in life.'"

Which brings us to today in Silicon Valley – where bold ideas are a norm.

What most distinguishes the culture of Silicon Valley is mental content and processing. Tech folks in general question and analyze. They practice focused thinking much or most of the time in contrast to their "normal" mind-wandering neighbors.

A minority among them, the Objectivists and Libertarians in particular, also know how to switch back and forth between analysis, synthesis, and visual imagination to pursue extreme "impossible" goals. They ask questions relentlessly. No idea is too extreme to at least look and consider as they search for better ways to do things.

In contrast, the mental content of most professional and business people is searching for and applying what's already known and practiced.

In further contrast, the mental content of those who have surrendered to conventional standards mainly live on a perceptual stream of consciousness.

At one time few would seriously start planning how to mine asteroids. Or pursue tourist travel to the moon and Mars. Or increase healthy human lifespan way beyond 100 years. Or provide free, superior education to all, world wide.

Now the "how-to" questions are being asked and visionary futures are on the way to being made real. Intellectual motivation toward solving the grand problems is much as promulgated by Peter Diamandis, Singularity University and the concept of Massive Transformative Purpose. Much MTP is now being applied by admirers of Ayn Rand and those influenced by her admirers as you'll see in the chapters ahead.

Meanwhile, when traditions are overthrown, so are repressive laws and regulations that have been enacted to enforce the status quo. So those who "think different" sometimes have to think and act like outlaws, an occupation well-suited to Objectivists and Libertarians.

CHAPTER FOUR   Breaking Bad Laws

One has a moral responsibility to disobey unjust laws.

Martin Luther King, Jr.

The outlaw factor is a large part of the Silicon Valley mystique.

Entrepreneurial innovation that bypasses business traditions will often go against government regulations.

It's analogous to the days when we had one phone company, guarded by law, with the insistence that Ma Bell was protecting the public interest by blocking competition. Similarly, there were three TV networks and three national airlines. Every interstate truck delivery had to be documented and approved by the federal government.

On the personal side, if you were black and sat at the wrong lunch counter, or gay and socialized with others in a bar, you risked arrest. Those who wanted to educate their own children did so under threat of fines and jail time (and some still are).

The status quo was required by law. Only when people figured out how to circumvent the law and pushed the issue by protest or openly breaking the law, did these realms become liberated. Still pending, as a current example: undoing the state electrical energy monopolies that control and maintain obsolete technology and high prices. Compare the U.S. with countries like Germany, where anyone can start an energy company and free-wheeling competition is the rule. (We have a similar freedom in Texas where you can have over 100 choices for electricity providers.)

"Silicon Valley" is synonymous with outlaw enterprise. But such enterprise is not geographically limited to the San Francisco/San Jose Bay Area. Outlaw enterprise is rife in Seattle, Los Angeles, New York, Boston, everywhere in between and worldwide to boot.

Bad laws are one issue. Repressive conservative attitudes are another. If anyone had proposed an smartphone-based ride-hailing system to a conventional corporate board of directors, for example, it wouldn't have gone beyond the board room. "It would be illegal." "The taxi drivers unions would block it." "Taxi permits cost a fortune and are strictly limited in most cities." And so forth and so on. Saying no to a new idea requires no thinking whatsoever, just a gut reaction and a thumbs down.

So there would be no Uber, no Lyft, and none of the other varieties now found in cities world wide. It's all illegal, vast, and almost everyone who uses such services are pretty happy compared to their experiences with their local regulated taxi service.

AirB&B has similarly challenged the hotel/motel industry with illegal room rentals. The hotel lobbies are gathering steam, paying off local legislators, and the battle is ongoing.

Pot sales, a crime for which some have been sentenced to life imprisonment, will now become a multi-billion dollar industry. Legalization, led by Objectivist, Marc Emery who, with his wife, did prison time during their campaign, is taking hold in state after state. So what was once a dismal back-alley business where black kids were busted for selling pot to white kids goes legit. Those were easy busts by racist police who filled our prisons and created a long-term social time bomb.

Alternative currency has long been illegal in the U.S. Now we have multiple varieties of alternative money including cryptocurrency such as BitCoin.

The true creator of the blockchain algorithm that makes BitCoin possible is much debated. The reclusive, most likely founder, Hal Finney, was remembered by fellow students for carrying around a worn copy of *Atlas Shrugged*. He was also remembered by classmates for always being ahead of his teachers. Meanwhile, a leading exponent of BitCoin, Nick Szabo gives credit to Ayn Rand for inspiration behind the concept.

Next on the legal agenda – laws to be broken and repealed:

- **Eliminating the asset forfeiture laws** that allow police to take your property if you're detained "on suspicion." You have to sue to get your property back even if you're released without charge. Hundreds of police departments are filling their coffers, and administrators are filling their pockets with this legalized theft.

- **Eliminating the 1970 War on Drugs** that still fills prisons with blacks and latinos. (FBI statistics show 80% of drug arrests are of blacks, 80% of drug use is among whites.) The "war" has caused massive corruption within U.S. law enforcement agencies and entire governments in Central and South America that serve the U.S. drug demand.

- **Eliminating government-sanctioned subprime student loans** that lock students into life-long unlimited interest charges and harassment by private companies that buy the loans. Refunds are virtually non-existent in higher education regardless of whether education is completed or of value. (It's a rerun of the subprime housing loan program that initiated the last great recession. A massive forfeiture of the multi-trillion dollar debt is overdue and may trigger another national collapse of financial markets).

- **Eliminating professional licensing laws** that protect no one but the licensed. (Similar to state laws that require you to buy your cars from dealers and pay their commissions rather than buy wholesale directly from the manufacturers. Elon Musk is working on it.)

- As mentioned previously: **elimination of state protected public utility monopolies** that block competition in energy markets.

What's the Silicon Valley connection? All these issues are being addressed by Objectivist and Libertarian activists and attorneys, many of whom are being financed by Silicon Valley entrepreneurs.

Protection of human rights is the only proper realm of law and regulation. When law runs counter to that, as in cases of fake deregulation and corrupted privatization, then there's another level to the problem. Cases in point being where judges' families own stock in private prison companies or judges take commissions for sentencing juveniles to private incarceration ("rehabilitation") facilities. Or privatized water and power companies are given monopoly licenses to risk their customers' lives – the worst of all worlds.

Skeptics raise the question: How do you judge what's a bad law when a majority of citizens or legislators think otherwise?

The only defensible purpose of law is to protect life, liberty, and property. It's predicated on the fact that the vast majority of people are perfectly capable of running their own lives and should be free to enjoy the fruits of their labor. Historically, the freer people are to learn, think, and interact, cooperate, and trade with others, the greater the wealth they generate. The greater the wealth, the better able they are to secure and improve their lives.

On that basis, laws that restrict access to knowledge, communication, travel, and trade are explicitly bad laws. Laws that impinge on private personal behavior are bad laws. Laws that discriminate and restrict the rights of members of a gender, ethnic group, or voluntary organization are bad laws. All such laws are ripe to be broken.

Some believe that their ancestral or ethnic origins, genetic, moral, or intellectual superiority grants them legal superiority over others. Such assumptions are grounded in the idea that some people are lesser beings to be ruled by the divine right of kings, or as in Nazi Germany, superiority derived from gene transfer from Nordic gods.

All people, unless physically/mentally impaired, have the reasoning skills and ability to support and improve themselves through free exchange with others. Billions of people around the world prove this daily, often under very difficult circumstances. What people seek and fight for endlessly are **good** laws, Libertarian laws that respect their right to support themselves and dependents, and to live their lives as they see fit.

CHAPTER FIVE   Thinking About Thinking

The things to do are the things that need doing, that you see need to be done, and that no one else seems to see need to be done.
                                                                Buckminster Fuller*

Ayn Rand held famed inventor Buckminster Fuller in high esteem. He identified the essence of invention as quoted above.

Anyone in any trade or profession can find plenty that can be improved or reinvented from the ground up. Or see wrongs that have been blindly accepted and find ways to fix them. This is revealed time and again in all professions. For example:

• Objectivist Robert Poole researched the economics of private versus public services throughout local governments. His writings sparked the best of the privatization projects that have been adopted worldwide. (No defense of private prisons here; as noted before, that's an enterprise that maximizes profits by maximizing imprisonments.)

• Objectivist Martin Anderson and Ayn Rand's attorney Henry Mark Holzer were active in the anti-draft movement. They were instrumental in persuading the Nixon administration to end it.

• Libertarian psychiatrist Thomas Szasz identified and publicized irrationalities in psychiatric law that allowed children and adults to be committed to mental institutions on flimsy pretenses. His work resulted in the liberation of thousands of falsely incarcerated asylum inmates.

** The things to do are the things that need doing, that **you** see need to be done, and that no one else seems to see need to be done. Then you will conceive your own way of doing that which needs to be done – that no one else has you do to or how to do it. This will bring out the most real you that often gets buried inside a character that has acquired a superficial array of behaviors induced or imposed by others on the individual.* From a letter to "Michael," February 16, 1970. Michael was a 10-year-old boy who had inquired in a letter as to whether Fuller was a "doer" or a "thinker." From *Critical Path* by Buckminster Fuller (2002).

- Objectivist Nathaniel Branden identified psychological causes of low self-esteem, a major cause of self-repression and underachievement. Authentic self-esteem is based on confidence in one's personal commitment to rational decision-making and behavior – not on the opinions and esteem of others.

One great secret to invention is questioning whatever realm that interests you: What is this for? Should it be done? What most needs improvement? What's most wrong with the status quo? What else would do it better? Those who follow this path often make an excellent living at it and, after the usual conservative resistance, become leaders in their fields.

This type of questioning is fundamental to Objectivism, and is a key to innovation: Find what's wrong, what isn't working, what needs doing – find the unmet and often unnoticed needs and figure out how to fill them.

One such unmet need is how to think better and faster. Finding new and better ways of figuring things out is central to Silicon Valley problem-solving methodologies.

A Silicon Valley hotbed of alternative thinking and problem solving is Singularity University, founded by Ray Kurzweil and Peter Diamandis. SU trains participants in "Exponential Thinking," "Massive Transformative Purpose," and "Design Thinking."

"Exponential Thinking" means looking at progress, not in linear step-by-step terms, which is our natural inclination, but in hyper-rapid acceleration.

In one description: Technology grows exponentially. Each advancement comes in half the time of the previous one. This is vastly faster than normal, linear thinking allows.

If you want to think exponentially, start with "Massive Transformative Purpose." That means that when setting goals, for example, you think in terms of the ultimate of what you might achieve in the extended future. Then you start working to make it real.

Massive Transformative Purpose (MTP) means going beyond the usual view of limited possibilities. In medical research, instead of thinking in terms of curing a single disease, for example, you think in terms of curing all diseases of aging to achieve indefinite life-spans. In education, you don't think of just improving present systems, you think about radical new paradigms that will accelerate learning vastly beyond today's norms.

"Design Thinking" is a multi-dimensional problem-solving structured approach to free-wheeling idea searches and testing methods.

Design Thinking includes "ideation," a form of extended brainstorming. Graphic analysis and idea-generating systems such as Mind-Mapping, Synectics, and creating Morphological Boxes to display all the possible options for solutions to any particular problem. (If these alternative thinking and problem-solving methods aren't familiar to you in your normal business operations, you'll better understand what makes Silicon Valley so unique.)

Here are some of the Massive Transformative Purposes now being tenaciously pursued by Objectivists and Libertarians in Silicon Valley and elsewhere:

- **Developing private space exploration, tourism, and asteroid mining.**

- **Extending human life-span without limits.**

- **Providing worldwide open access to information and education.**

- **Creating universal environmental sustainability and pollution-free energy.**

- **Overthrowing all dictators and authoritarian governments.**

- **Enhancing the human brain and vastly expanding artificial intelligence.**

In each case, they're looking at what's in the way. What has to be eliminated to achieve a goal? Elon Musk, for example, asked what was preventing profitable space travel. The answer – the financial stopping point – was throwing booster rockets into the ocean instead of reverse landing and reusing them.

CHAPTER SIX   Private Space Exploration, Investment, and Development

I think fundamentally the future is vastly more exciting and interesting if we're a spacefaring civilization and a multiplanet species than if we're not.
Elon Musk

Next on the private space enterprise agenda:

- **Lunar tourism and colonization.**
- **Mars, ditto.**
- **Mining the asteroids.**

Steve Davis, SpaceX engineer introduced as "employee #14," spoke at a Washington DC, Objectivist conference in the summer of 2012. Steve is well familiar with *Atlas Shrugged*, quotes from it, and declared that his boss, Elon Musk, "has read the book."

Musk is moving along on his ambitions to establish lunar and Martian colonies. If he can continue the momentum of the recent spectacular Falcon rocket launches, he has an excellent chance of making it. He comments: "You want to be inspired by things. You want to wake up in the morning and think the future is going to be great. And that's what being a spacefaring civilization is all about."

Peter Diamandis was an Ayn Rand enthusiast since his early days getting dual degrees in medicine and astronautics at Harvard and MIT. He founded Space University, the Space X Lunar Prize, Planetary Resources, and other enterprises related to AI, education, and life extension. He has been a spark plug for other entrepreneurs who are driving private space development such as Richard Branson's Virgin Galactic.

Bryan Johnson sold his company Braintree Payment Solutions to eBay for $800 million and is now using his funds for space development and brain enhancement research. He is also a Director at Human Longevity, Inc.

Bryan says: "Atlas Shrugged highlights the beauty of capitalism and the dangers of government intervention." He has teamed with Google's Larry Page to develop the Planetary Resources company to mine resources from asteroids and other planets.

Planetary Resources recently successfully launched its Arkyd-6 CubeSat, a prototype satellite for prospecting minerals and water content of asteroids.

In an article on *Medium*, Johnson wrote:

"We are at one of the most exciting moments in history. At no other time has the distance between imagination and creation been so narrow. We now have the power to build the kind of world we could previously only dream of. With new tools such as 3D printing, genomics, machine intelligence, software, synthetic biology and others, we can now make in days, weeks, or months, things that previous innovators couldn't possibly create in a lifetime. Where da Vinci could sketch, today we can build. And yet, there are still so many problems that we haven't begun to solve, so many rich opportunities that lie in wait."

Another space Libertarian, Jim Cantrell is a well-known entrepreneur and expert in space systems, with 30 years of experience in the aerospace and high technology industries. Jim was part of the SpaceX founding team and now founder of Vector Space Systems to launch low-cost satellite deployment. This may blend well with others' plans to place thousands of communications satellites to communicate with and provide education for billions of people who are still out of range of the internet.

While much is made of the possibilities of space colonization, the big money would be in mining precious minerals in asteroids.

Long-time Libertarian, Amara Graps, is a physicist pioneer in asteroid mining science. She led a 2016 conference in Luxemburg, creating a ground-breaking roadmap toward asteroid development: "Asteroid Science Intersections With In-Space Mine Engineering (ASIME)." Her most recent, 2018 ASIME conference in Luxemburg and the science that was explored, may trigger a surge in international investment in asteroid prospecting.

Current estimates of the wealth available in the Asteroid Belt are, well, astronomical – multi-trillions of dollars in revenue. Hayden Planetarium director Neil deGrasse Tyson predicts that whoever "first learns to mine asteroids and bring back their minerals to Earth will become the first trillionaire. There's no question about that in my mind."

CHAPTER SEVEN   Life Extension

Self-transformation is more than a mere preference. Ayn Rand identified the fundamental and ultimate choice perpetually confronting every organism: To live or to die.
                                            Max More, CEO of the Alcor Life Extension Foundation

Life extension used to be the province of health diets and supplements. Now the term refers to goals vastly beyond what conventional medical science previously considered possible or even desirable.

First in line were Libertarian activists Durk Pearson and Sandy Shaw with their best-selling 1982 book *Life Extension*. Although their emphasis on multiple antioxidant supplements has been much questioned, these MIT and UCLA graduates sparked widespread awareness of the possibility of healthy extended longevity. They also won a court case they filed against the FDA regarding government regulation of supplement advertising.

Libertarian Sonia Arrison, Senior Fellow of the free-market Pacific Research Institute and a Trustee of Singularity University, wrote the book *100+ on the coming age of longevity.* She advised Peter Thiel on the subject and has also studied and written on a number of topics related to accelerating technology and human enhancement.

Much current Objectivist/Libertarian financed research is aimed at carrying this concept further than Ayn Rand ever envisioned.

Max More's ALCOR, incidentally, recently received Bitcoin pioneer (and likely originator), Hal Finney's body for cryogenic preparation for later resurrection.

In the realm of genetic research on longevity, Ayn Rand admirer Peter Diamandis, has teamed up with Craig Venter, the first to map the human genome, to create Human Longevity Inc. He most recently cofounded a related huge new research enterprise, Celularity.

Add to this mix, Peter Thiel's investment in cutting-edge longevity research and Google's semi-secret Calico longevity research project. There's a long and growing list of other investments in AgeX, SENs, Buck Institute, and many others. It's enough to to suggest the aging problem will be solved within the lifetimes of many us.

A new book – *Longevity: The State of the Art and Science of Cheating Death* by an Objectivist educational organization (www.lifexedu.org), will be released in summer of 2018. It covers all of today's best scientific work on life extension with the intention of informing and attracting medical professionals to this realm of research and practice.

One notable medical professional is Dr. Shira Miller, who became interested in Ayn Rand's writing in college. She says Objectivism provides her with "a moral philosophy for how to live a good and happy life." Dr. Miller practices age-management medicine in Los Angeles and is a pioneer in age-related therapies. Many other medical professionals will be coming on board soon.

Many of those searching for methods of self-enhancement belonged to the Extropian Society years ago, which has now morphed into Transhumanism (which includes many Libertarians and Objectivists as members).

Hal Finney, cited above, was described by BitCoin promulgator Nick Szabo as a "frequent participant in the Extropian online discussions of cryonics, life extension, space colonization, nanotechnology, artificial intelligence, mind uploading (the transfer of human consciousness from biological brains to alternative high-performance substrates), and the technical, social and political aspects of cryptography. Julian Assange 'lurked,' and occasionally participated."

Typical of the anti-death mindset, Oracle founder Larry Ellison, told his biographer "Death has never made any sense to me. . . . How can a person be there and then just vanish, just not be there?" Ellison is a major player in longevity research. According to some reports, he's donated more than $430 million to the cause.

CHAPTER EIGHT   Liberating Education, Part One – the Early Years

The State shall make no law regarding the establishment of education.
U.S. constitutional amendment
proposed by educator ands scholar Ivan Illich

Ayn Rand once called public schools the world's largest concentration camp. Political writer Paul Goodman echoed Rand, describing public schools as "an arm of the police, providing cops and concentration camps."

Wherever people are forced by law to be confined and controlled, that would qualify as a jail or concentration camp. And that's what most elementary and high school children face in their daily lives for most of every school year.

As for learning, roughly one-third of each school year is devoted to reviewing the content of the previous year, and another third is intended to preview the next year. New content is supposed to be plopped somewhere in the middle.

The content of the grades from 1 to 6 can be learned in one year. This has been proven on occasion when an adolescent child is discovered who has skipped all standard schooling and does a quick catchup. The standard reading, writing, and math skills conveyed up to high school require about a year of part-time study. High school skills take longer but are often less than beneficial. The last chapter of most high school math textbooks, for example, is usually barely distinguishable from the first chapter.

Education reformers have raised all these points for generations, but there are few changes in the public schools other than federally mandated tests and more tests, all of which are accompanied by less learning and lower scores.

After generations of "Why Johnny Can't Read" and related books, conferences, and school board meetings, there's no reason to expect positive changes in public elementary or high schools.

There are more ways to escape the system now for those families who understand their children's innate desire to learn. Montessori schools, for example, offer the youngest students self-guided and self-paced learning projects. Some, such as the Fountainhead Montessori schools in the San Francisco Bay Area, are managed by Objectivists.

Homeschooling is a favorite of Libertarian/Objectivist parents (as well as religious parents to be sure). While under constant attack by government officials, homeschool parents have a national legal defense system that generally succeeds in protecting childrens' right to be truants.

One self-educated child who later taught at UC Berkeley realized that elementary school state funding depended on recorded attendance or signed absence excuse notes. So he forged enough notes to salvage one or two days a week out of school when he could go to a public library and learn what he wanted. The library had a children's section and children weren't allowed to check out books from the adult section. So our friend learned how to hide and smuggle the books he needed most out of the library (including *The Fountainhead* and *Atlas Shrugged*).

Students have figured out how to game the system by using Kahn's Academy for online "instant learning" videos on basic school topics. That opens up time to pursue learning that's most important to them. Besides using Kahn's videos for basic school requirements, Wikipedia and Google are virtually unlimited sources of information for whatever it is that anyone wants to learn. (Salman Kahn, CEO of Khan Academy, describes himself as a capitalist educator. He just earned the 2018 Visionary of the Year award in San Francisco.)

None of this is to say that the public school system isn't okay for many students who enjoy the socializing, their clique or clubs, phys ed, etc. and get by just fine. They're likely to join the path laid out for "normal" folks: College, a reasonably well-paying job, marriage, kids, and then retire, travel, socialize, and play golf. What's not to like?

What's missing is an alternative path for independent-minded students who want to focus and accelerate their education on their own terms.

This is what was proposed and predicted by educationalists Paul Goodman and Ivan Illich in the 1960's and 70's. (Paul Goodman and Ivan Illich were pre-libertarians who have been powerful influences on the experimental educationalists in Silicon Valley.)

Paul Goodman proposed in *Compulsory-Miseducation*, a step away from universal public education: "that, on the model of the GI-Bill, we experiment, giving the school money directly to the high-school-age adolescents, for any plausible self-chosen education proposals, such as purposeful travel or individual enterprise. This would also, of course, lead to the proliferation of experimental schools." He also advocated more liberal teacher credentialing: "Use appropriate unlicensed adults of the community – the druggist, the storekeeper, the mechanic – as the proper education of the young into the grown-up world."

As an advocate of nonviolent resistance, Paul Goodman described what he called "A Touchstone for the Libertarian Program" "The Touchstone is this," he writes, "does our program involve a large number of precisely those acts and words for which persons are in fact thrown into jail?" This is what happened during the Civil Rights movement in general and the widespread confrontation with police during the Gay Rights movement. And indeed, people are still being arrested for refusing to obey local school attendance laws.

Not surprisingly, Silicon Valley has become a haven for alternative schools and many of the entrepreneurs named in this report are investors. AltSchool, Montessori, Waldorph, Ad Astra, Sudbury, Kahn Lab, all have their attractions. Most unique perhaps is the Kahn Lab where students are encouraged to study their education and find ways to improve their school. Many alternative schools incorporate the most obvious needs, such as open access to study materials, elimination of the annual progression through grade levels, democratic student participation, but they're still based on the classroom/teacher model.

A longstanding model for alternative schools is the Bing nursery school on the Stanford University campus. The core curriculum is free play, which in the case of young children means freedom to interact with reality and learn from it. But annual tuition of $20,000 or more to pay for classrooms, teachers, teaching materials, etc., are a barrier for most of the U.S., not to mention the world's population.

Paul Goodman proposes something more appropriate to the age of the internet where all knowledge is available free online and hundreds of thousands of volunteers would offer their services as tutors on demand. "Exchanges needed to be set up where a person could find a teacher for what he or she wanted to know. Education cooperatives, where students were matched with teachers, would replace schools, where bureaucrats possessed the authority to define learning for everyone else. And everyone, no matter their age or certification, would be both student and teacher."

Despite the lack of reform among professional educators, there is no shortage of classic evidence-based literature on the subject such as: Peter Gray, Ph.D.: *Free to Learn* and John Holt: *How Children Learn* and *How Children Fail*

CHAPTER NINE   Liberating Education, Part Two – Higher.Edu

We who take our education outside and beyond the classroom understand how actions build a better world. We will change the world regardless of the letters after our names.
Dale Stephens,
Thiel Fellowship recipient, founder of UnCollege

Higher education has problems that universities have been reluctant to own up to:

• The traditional path of education and career has been narrow, restrictive, and oppressive. That pipeline model channels students through a college system that hasn't changed in over 150 years.

• Most people who need higher education are working professionals who can't afford the time and money required for campus-based degree programs.

• There's limited access to experts and courses. Every campus has its experts but that doesn't allow students to have access to the hundreds more at other campuses. Taking multiple courses at different schools simultaneously is prohibited by most universities.

• No credit for self-education. You might learn more, but if you're not paying for it on campus, it doesn't count.

• An archaic system of accreditation where peers judge peers to verify that they're all following the same obsolete traditions.

• Time-consuming inefficiencies. For example, the old model of education has students sit through a lecture for 50 minutes, inputting notes from the professor's notes (that could be read online in 10 minutes). This huge time waste just doesn't cut it anymore. As for interaction with faculty, that's rare except in graduate seminars.

• Overpriced tuition that pays for non-academic campus amenities and inflated academic salaries. It's preposterously expensive thanks to the inflation factor of the scandal-ridden federal student loan programs.

The costs of running online education are a small fraction of on-campus education, but most schools still charge full fees for their online programs. Many online programs follow the old model of lectures, readings, and cramming for exams – a waste of the potential of internet resources.

Administrators feel they must do this to avoid diverting income from their standard enrollment tuition. And faculty are, of course, worried about losing their jobs. So the age-old systems and exorbitant tuitions common to campus-based schools remain unchanged.

These are all factors that were observed by faculty members at Stanford University who founded the Massive Open Online Courses (MOOC) of Udacity and Coursera. These programs require considerable self-motivation and Ishan Gupta, Managing Director at Udacity says: "One book to read when you want to motivate yourself is *The Fountainhead* by Ayn Rand."

Traditional campus-based schools fill a need for those students still undecided as to what they want to do in life. It's a chance to test the waters. It's a great chance to party every Friday. It also serves the social and financial needs of those who can afford the tuition and have a family tradition of membership among the Harvard, Yale, Princeton, Stanford, etc. alumni elite.

Students of any age who know what they want to learn will learn it no matter what. They just need reliable direction to the best online sources, low tuition (if any), and, if required, some objective evidence they can display to show they've earned academic credit, certificates, and degrees.

The leading resource for all students, credit or no credit, is Wikipedia. Wikipedia is the go-to place for most online inquiries and information. The ultimate goal for its co-founder, Objectivist Jimmy Wales, is no less than free access to all the world's knowledge. They're getting there.

Wales has shied from publicity, but the impact of his work on education has been compared to that of Steve Jobs. Jobs provided the media, Google is providing the search engine, and Wales and his 80,000 member army of volunteers are providing the content.

Wales observes: "When you think of the university-aged kids now, they haven't known a world without Wikipedia, to them we are part of the infrastructure of the world."

Schools have had to scramble and have not been fully successful in finding faculty competent to teach artificial intelligence, robotics, machine learning, blockchain technology, life extension geriatrics, and tech-based enterprise. Many entrepreneurs who are now leaders in their fields had to drop out of school in order to learn what they needed.

Major expansions in media access to provide free worldwide education are currently planned by Objectivist and Libertarian members of what some educators have called the "Silicon Valley Mafia." To meet the need, there are new opportunities in entrepreneurial technology education education sources including:

• Objectivist Carter Laren is a director of Founders Institute that operates an international school for entrepreneurs. They operate an intensive sequential question-focused curriculum that any Objectivist would admire.

• In a similar vein, Jawed Karim a cofounder of YouTube also founded Youniversity Ventures to advise and help current and former university students create and launch new business concepts.

• Singularity University, cofounded by Peter Diamandis, follows a similar tact. Their students sign up, go to work asking questions about problems that need solutions, find answers, and make things happen.

• An Objectivist-funded initiative is Futurize X at UCLA. It invites students to focus on accelerated technologies such as AI, Space Development, Longevity and Environmental issues.

• Coursera and Udacity are two examples of ambitious outreach with many thousands of students at a time in AI education, for example. But solving the problem of reliable validation and granting credit for the learning experience remains, especially for those seeking to transfer credit from school to school.

There is no shortage of knowledge. Anyone can go online and learn almost anything about anything. If book knowledge is required, that too is often now available at very low, or no cost. Skills are another matter, but folks are working on it.

Meanwhile, forward-looking companies such as Google have statistically documented what's long been common knowledge: Diplomas have no bearing on employee performance. Degrees will no longer be a requirement for employment at Google.

For those seeking a transcript of on-site participation in advanced technology there are also a number of programs shown below.

The Objectivist-based San Francisco Institute of Architecture's online programs, for example, require students to provide "tell-backs" summarizing what's learned from the textbooks and other sources. This is more effective and efficient than the lecture-cram-exam method which has a very low information retention rate.

The "tell-back" principle is simple: reading is far faster for gathering information than listening to lectures. Reading allows a quick review to clarify anything not fully understood, and summarizing what's learned in your own words integrates the information in one's brain far better than cramming for exams.

There are values in contact with truly expert faculty but one of the pitfalls of campus-based education is that the experts are scattered all over the world. Any particular campus is limited in resources. Actual average learning time in any curriculum is quite limited – some 20+ hours a week according to the Bureau of Labor Statistics.

A potential online student population in the billions points to enormous investment opportunities even for the lowest cost online distance learning services. The schools just need an amplified international satellite internet reach and an objective worldwide system of student accreditation.

Libertarian Peter Thiel financed the work of 20-year-old dropout, Dale J. Stephens and his concept of the UnCollege. As described by Stephens:

"The UnCollege 'Gap Year Program' is a 9-month program designed to equip young adults with the necessary skills to succeed in the innovation economy. . . . The program is divided into phases including an immersive service-learning experience abroad, and a skill-building module in San Francisco."

UnCollege tuition is $16,000, which includes housing and most meals but not air travel. Some student aid and payment plans are available. Best of all there are no tests, SAT scores, transcripts, or letters of recommendation required to enroll. Perfect for the dropout or those who have been forced out of traditional programs.

UnCollege provides a model for future pre-college or non-college education programs. Except for the cost of the international travel phase of the program, a student could pretty much assemble and manage their own personal program with no tuition.

More details from Stephens after he was awarded the grant from Peter Thiel: "I have been awarded a golden ticket to the heart of Silicon Valley: the Thiel Fellowship. The catch? For two years, I cannot be enrolled as a full-time student at an academic institution. For me, that's not an issue; I believe higher education is broken.

"I left college two months ago because it rewards conformity rather than independence, competition rather than collaboration, regurgitation rather than learning, and theory rather than application. Our creativity, innovation, and curiosity are schooled out of us.

"Failure is punished instead of seen as a learning opportunity. We think of college as a stepping-stone to success rather than a means to gain knowledge. College fails to empower us with the skills necessary to become productive members of today's global entrepreneurial economy.

"College is expensive. The College Board Policy Center found that the cost of public university tuition is about 3.6 times higher today than it was 30 years ago, adjusted for inflation. In the book '*Academically Adrift*,' sociology professors Richard Arum and Josipa Roksa say that 36% of college graduates showed no improvement in critical thinking, complex reasoning, or writing after four years of college. Student loan debt in the United States, unforgivable in the case of bankruptcy, outpaced credit card debt in 2010 and will top $1 trillion in 2011."

Stephens received scathing, insulting tweets as his remarks became public. This indicates how entrenched is the higher education status quo mindset. As the world changes, accredited, free, worldwide higher education will be attacked vociferously and probably outlawed in some jurisdictions. But the need is too great and the people behind it, such as Stephens and others like him, are too smart not to make it happen.

CHAPTER TEN   Environmental Sustainability.

The faster we achieve sustainability, the better. . . we want to change the entire energy infrastructure of the world to zero carbon.
Elon Musk

It's not well known, but buildings are the number-one consumers of electrical energy (around 70%). Buildings and related industries are the primary consumers of natural resources, the number one sources of pollution and greenhouse gases, and are disease-breeding domains of toxic materials and flawed air handling systems. (From statistical analysis from the U.S. Department of Energy.)

The solutions to these ills save enormous amounts of money. (This opportunity is mainly opposed by political conservatives who confuse environmental concerns with a left-wing conspiracy.)

The Objectivist/Libertarian-based school, San Francisco Institute of Architecture (SFIA) produced the first international "Eco Wave" Ecological Design conferences, and published the first textbook on the subject, *The Ecological Design Handbook*, through McGraw-Hill. The school is now all online as the world's largest school of architecture, green building, and sustainable design. (www.sfia.net.)

Jeffrey Skolls, cofounder of eBay and as noted before, much inspired by Ayn Rand, is a leading private funder of environmental causes through the Skoll Global Threats Fund. The fund assists over 60 non-profit foundations worldwide such as the Climate Protection Campaign. (Most advances in climate and environmental repair are, and will be, created by private sources and institutions.)

Peter Thiel believes we need to find our way back to far more fundamental visions and that we should be working toward a future of "radical breakthroughs," with things like "clean energy sources" and "deserts that can be transformed into fertile landscapes."

Elon Musk's stated goals include manufacturing lower cost batteries to better store electricity generated by solar panels, including an easy to manage storage system and totally off-the-grid solar power system for buildings.

Musk's Massive Transformative Purpose for Tesla and SolarCity is to accelerate the world's transition to sustainable energy. He says: "We must at some point achieve a sustainable energy economy or we will run out of fossil fuels to burn and civilization will collapse. Given that we must get off fossil fuels anyway and that virtually all scientists agree that dramatically increasing atmospheric and oceanic carbon levels is insane, the faster we achieve sustainability, the better."

Musk's plan: "Create a smoothly integrated and beautiful solar-roof-with-battery product that just works, empowering the individual as their own utility, and then scale that throughout the world. One ordering experience, one installation, one service contact, one phone app. And, of course, Tesla cars powered by the homeowner's solar power systems."

The universal goal among the giants of the industry is to make all their operations 100% powered by alternative energy sources. That's the center of attention at Silicon Valley conferences such as the Oracle annual Sustainability Summit events. It's also a big deal for other larger companies on our Objectivist/Libertarian founder Who's Who list in Chapter 2. And there are collaborations to share data among the concerned companies: Sustainable Silicon Valley, the Silicon Leadership Group, and others. The overriding goals are to produce more clean energy than consumed, sequester carbon, recycle water, and eliminate environmental toxins.

Government intrusions, public utility companies, and petroleum and coal industry lobbying will slow the transition to green building. But Objectivist and Libertarian architects, educators, entrepreneurs, and lawyers are among the prime movers in circumventing bad laws and regulation. A hint of things to come and already available: After a few years of payback time, homeowners already have free electrical energy to power their homes and cars at virtually no cost for the rest of their lives.

CHAPTER ELEVEN   Overthrowing Dictators and Authoritarian Governments

> *... even totalitarian dictatorships are dependent on the population and the societies they rule.*
> Gene Sharp

Libertarians responded well to Gene Sharp, today's leading advocate of non-violent resistance. Sharp worked at Harvard and the University of Massachusetts where he founded the Albert Einstein Institution. His words, quoted above, reflect Ayn Rand's observations regarding the "sanction of the victim" – that exploiters have no power other than that granted to them by the exploited.

Sharp's historical research has shown that time and time again, nonviolent resistance is more effective than violent rebellion in bringing about social and political reform.

This also correlates with Rand's dictum of the non-initiation of force as the basis of the proper relationship between government and its citizens. A signed agreement with this principle of restriction on government was a long-standing requirement for membership in the Libertarian Party.

Sharp's writings, such as *From Dictatorship to Democracy: A Conceptual Framework for Liberation,* were political handbooks for activists in the Arab Spring of 2010-2011. Nonviolent action was quite effective for the overthrow of repressive regimes. But in many cases, such as today's Egypt, it didn't prevent the subsequent overthrow of a start-up democracy by a new authoritarian regime. An effective way for people to secure freshly won democratic reforms remains a work in progress.

Liberty International was founded in 1989 and had remarkable impact in liberating the "soft" authoritarian states in Europe. At that time virtually every Western European nation had only two TV stations – one station run by the government and one by labor unions. Tax on net income in some Scandinavian countries went over 100% and much housing and industry was state-owned or managed. Their economies survived to the degree they still allowed some amount of free enterprise.

Libertarian International pioneered another revolution in disseminating pro-freedom literature. Disguised truckloads of books were sent to the Eastern European countries, again helping set the groundwork for defeating dictatorships and establishing democracies.

Totalitarian regimes once had an iron grip on impoverished Eastern Europe. Today, it's a vastly different world across Europe in large part thanks to Libertarians who became politically active and successfully explained and argued for free-market reforms.

The battle never ends, however, and Liberty International is going strong, especially with youth outreach.

To quote their web page: *LI is advancing world liberty through international education and networking including 33 World Conferences, including hundreds of student scholarships ... Liberty Camps that have taught principles of liberty and entrepreneurship to 5000 students in 30+ countries ... Liberty Book Translations in over 50 languages ... Eight annual China Austrian Economics Summer Schools for nearly 1000 students ... Nearly 5 million print and electronic issue papers distributed throughout the world.*

Although the international organizations are titled as Libertarian or Free Market, virtually everyone involved was influenced by Ayn Rand.

Students for Liberty is an Objectivist-based student organization that hosts over 1,000 university student chapters around the world. Seminars on Ayn Rand are standard fare at their conventions, and copies of *Atlas Shrugged* have been freely distributed.

These are the students who will work in think tanks, universities, and government administration positions worldwide. Their impact will be felt for generations to come.

Hopefully democratic institutions will also survive and thrive during economic reform, but that's not guaranteed.

In the cause of freedom, more than a generation ago, the Soviet Union experienced the consequences of the Samizdat Movement. As it happened, when the Soviet government confiscated all the means of production as property of the state, typewriters remained personal property. You couldn't make photocopies without a government permit, but you could type multiple carbon copies on typewriters. Thus did pro freedom literature become distributed throughout the Soviet Union and help lay the groundwork for dissolution of the Soviet empire. (Now we have the second-stage problem of once liberated countries spiraling back into dictatorships – the next challenge.)

Putin's former chief economic advisor, Andrei Illarionov, was an advocate of Objectivism and sought to have *Atlas Shrugged* distributed throughout the Russian school system. That didn't take, but he is now a Fellow in the Center for Global Liberty and Prosperity. A sister organization, the Center on Global Prosperity is a branch of the Independent Institute, the Bay Area Libertarian think tank long-supported by Peter Thiel.

The Moscow Times quoted Illarionov: "'Every import tariff and every limit on foreign-exchange transactions is a blow to our consciousness. Every tax acts against our freedom,' he said at a news conference dedicated to the launch of Ayn Rand's work in the Russian language."

Now, with the rise of electronic censorship by China, pro-freedom movements need more alternatives of disseminating information. If the Chinese freedom movement is fully blocked in online communication, it may have to revert to effective but risky underground book distribution.

Another hazard is the problem of U.S. intervention and "nation building."

U.S. government interventions in other nation's governments have been disastrous. For example, the deposing of the elected president of Iran and installation of the Shah of Iran monarchy by the U.S. (at the behest of the petroleum industry worried by a threat of nationalization of oil production) led to the Iranian "hostage crisis" that dominated the daily TV news for a year and cost President Carter his re-election.

The CIA attempted to block democratic and socialist movements in Central America by importing cocaine from Columbia and using profits from drug sales in U.S. cities to fund the Contra army. The results profited the drug cartels, established permanent routes of drugs into the U.S., and added to addiction problems in the poorest (black) U.S. urban areas.

U.S. intervention in Vietnam needs no further description except to say over 60,000 Americans and nearly 1 million Vietnamese died – all for absolutely nothing.

Ayn Rand and the Libertarian movement opposed U.S. intervention into other nation's affairs – stop funding dictatorships worldwide and leave pro-democracy movements to the volunteers. Results won't always be fast or great, but unexpected negative results won't be as massive and tragic as when it's backed and financed by the vast resources of the U.S. government.

There are other hazards such as when Communists become capitalists through liberation (more or less) of economics in China.

Mao's Red Guard of the Chinese Cultural Revolution in the mid-1960's destroyed libraries at Beijing University and books had to be replaced. After the Cultural Revolution faded, a call went out from a Libertarian educator at Semesters at Sea for donations of free-market economic textbooks to restock the University.

Books were sent to Semesters at Sea mainly from members of the Libertarian International and their ship delivered much-welcomed textbooks to China.

Libertarians and Objectivists involved in this effort joked at the time that China might become the first free-market Communist state.

That's what has happened. Although corruption and cronyism are rife, business went private and China's economy has expanded exponentially. It's to the point where you would be unlikely to find any item at Walmart or Target or your local drug store that wasn't made in China. Now that President Xi has established himself as the new president for life, Chinese investments will increasingly find their way overseas, as will the Chinese business elite who can afford to buy citizenship in the U.S. or Canada.

As the Communist Party cracks down more intensely on communication and the flow of ideas, the best Chinese minds will go on strike or depart, the country's prosperity will decline, and "interference by the West" will be the scapegoat, as it is now in Venezuela. As predicted by *Forbes:* "The economic impact of this change (in China) will likely range from negative to disastrous." That, in turn, will lead to further crackdowns, more political prisoners, and an ultimate breakdown. Then we'll see how effective the currently low-profile, pro-freedom organizations and student groups will be.

On other fronts, Objectivist Jeff Skoll's Foundation, in addition to fighting global climate change, has an international liberation role with particular emphasis on democratization in the Middle East.

Still another movement to help relieve poverty and liberate economics in the Third World is "Kiva." Kiva is a nonprofit channel for providing loans to third-world entrepreneurs

When looking at the problem of poverty in the Third World, a major reason people can't better themselves is that they don't own the land their ancestors have lived on for generations. They don't have titles, and can be booted off at any time by government or absent landowner whim. So they don't have collateral to borrow to invest in taking steps to better themselves financially. Liberation programs are being introduced worldwide now to deal with this problem.

In a related vein, loans to small startups are almost impossible to find in conventional banking sources in the Third World. Most such banks require a year's estimated income as collateral for loans. Several programs, such as Kiva, have been founded to provide loans on a more realistic basis, funded by volunteers.

An Objectivist, Andrew Cooper, is a typical supporter because, as he says, such programs circumvent government cronyism and bureaucracy. He is reported as having "discovered *Atlas Shrugged"* in college after reading Milton Friedman's *Free to Choose*. He considers *Atlas* to be "one of the most profound books" he has read. He credits Rand's writing to his understanding that "big government is not only ineffective and counterproductive but manifestly unethical."

CHAPTER TWELVE   Enhancing the Human Brain, Expanding Artificial Intelligence

One estimate is that the brain can perform 38 thousand trillion operations per second. A supercomputer can manage .002% of that.
Quora Digest

Atlas Shrugged enthusiast Bryan Johnson is the founder and CEO of Kernel, a company developing a neuro prosthetic device to improve brain function. Johnson founded Kernel in 2016, investing $100M of his personal capital to build advanced neural interfaces to treat disease and dysfunction, illuminate the mechanisms of intelligence, and extend cognition.

The goal is to create a device to improve communication between brain cells by hacking the "neural code." If properly tuned, this could send signals to mend a cognitive dysfunction.

"We have done this before with biology and genomics," said Johnson, "We can program yeast to do a specific function. We can expect the same path with neural code."

Those who face cognitive deficiencies will, according to Johnson, be the first super-humans. People will be able to purchase intelligence and provide self-care for anxiety and depression. Johnson's intent, he says, "is to build products for billions of people, not just the elite or rich."

Neural prosthetics could also provide brain and artificial intelligence (AI) interfaces to enhance the capabilities of both. In the realm of artificial intelligence, Objectivist AI researcher and entrepreneur, Peter Voss, notes that most who work in AI have been at a disadvantage because they're not aware of the Objectivist theory of knowledge and consciousness.

The model of human consciousness offered by Objectivism points the way to enhancing human mental acuity. The main point being to learn how to activate the dominant organ of human intelligence, the lower frontal cortex of the brain, specifically the little known but most vital part of the human organism, the Orbital Cortex.

Note that the brain is not a single organ. It's a combination of organs that work in symbiotic collaboration (most of the time, in most people). As we all know, sometimes coordination goes haywire and people become mentally dysfunctional.

The level of neural organization in the human brain is beyond comprehension, with about 90 billion brain cells (neurons) with sometimes as many as hundreds of neural connectors (axons and dendrites) within each neuron. The total neural communication can number in the thousands of trillions, per second as noted previously.

That's why we're so smart, although a large part of our mental energy is applied to self-repression of our innate, almost limitless intelligence.

For example, those who hold religious beliefs that purport to answer all important questions will not pursue further questions. Those who believe their thoughts drift into their heads from a higher reality won't attempt to manage their mental processes. Those who haven't learned how to think clearly, live in chronic fear of the unknown and rely on others to do their thinking for them.

The Orbital Cortex of our brain is not shared by any other mammals. It's not large, about the size of a silver dollar, but it's the chief administrator of all our thinking. It's where you ask questions, formulate answers, identify solutions, weigh options, and formulate plans. It's our rational self.

The Orbital Cortex, the most human of human organs is unmatched in complexity and capability by any other mammals, regardless of their overall brain sizes. (For more detail see *The Executive Brain: Frontal Lobes and the Civilized Mind* by Elkhonon Goldberg.)

To think clearly and maximize thought power, you have to know how to switch on the appropriate brain lobes and maximize the number of neurons working for you – maximize your consciousness in other words.

One direction that Objectivist researchers are taking is to create a device that allows you to see what parts of your brain are lighting up at any one time. That way you can shift focus to the part of your brain most competent at dealing with whatever you want to be thinking about. In some cases such as visual problem-solving modeling, it will be the visual cortex at the back of the brain. In other situations, it might be focus on the left or right hemispheres.

There'll be more details on this in a later chapter on education that describes research on learning at DARPA, Johns Hopkins, and other institutions that are studying the enhancement of learning stimulated by external electrical brain stimulation.

During Ayn Rand's lectures, she stated that the one choice we have, the one realm where we have volition, is "to think or not to think."

A questioner would usually ask: "But how do we choose to think?" Rand's reply was: "How did you choose to ask that question?" That's how simple it is, like lifting a finger or turning your head, you're not aware of all the nerve connections involved. Like the *Nike* slogan, you "just do it."

There's no question, however, that having a device that shows you visually where your neurons are firing will help provide a new measure of control over mental processes.

Want to avoid expressing a negative emotional outburst? There's a place (or two) in the brain for that. Want to stop ruminating and get to sleep? Theres a place for that too. Want to stop procrastinating an important but unwanted task? Sort out an internal conflict and make a rational choice? It's all there. It's **your** brain after all, you just need to see and experience it clearly enough to take charge and activate the lobes appropriate to whatever it is you want to do.

The brain's collection of specialized organs work in multiple interactions that create what we experience as consciousness, memory, emotions, and thought.

Brain content starts with the Cerebellum, near the top of the spinal column, that receives incoming sensory data (sensates) and routes the raw input, our direct experience of the world around us. Raw visual input is the most complex, and is transmitted through the optic nerves to the Occipital lobe, also known as the visual cortex at the back of the brain. That's what you see in the instant before you translate the visual image of this page into words and concepts.

Repeated sensory inputs leave neurological traces, become recognized with repetition and are then named. The combined and integrated sensates are "percepts."

For example, a child sees mom and dad and hears sounds linked to them ("mama" and "daddy"). The experiences are repeated and neurologically linked over time. The linked experiences and words are routed through the next step up the ladder of consciousness, the Parietal Lobe, and Cerebrum. (There is also routing through the Thalamus, Hippocampus, and Amygdala in the center of the brain. They provide other functions such as directing memory storage, growing new neurons, and triggering the hormonal reactions we experience and name as emotions.)

Just as sensates become linked with symbols, such as words, and experienced as percepts, so are percepts linked and grouped by similarities, then named and retained as abstract concepts. Abstract concepts are content for the frontal lobes, the top realm of human consciousness.

The human species evolved by expanding operations of categorization, induction deduction, hypothesizing, etc. in the Prefrontal Cortex. These lead to increasingly useful abstract concepts such as cause and effect, reasoning, logic, and problem-solving. On the other hand, most day-to-day functioning is routine and competently handled on the perceptual level (the Cerebrum). But foreseeing events, planning, and problem-solving are abstract functions reserved to the Prefrontal Cortex and especially the Orbital.

That's where we structure thought by formulating questions. You can test this any time you catch yourself doing associative mind wandering. Just pause and refocus on your lower frontal cortex behind your forehead and question the importance or relevance of what you're doing. That will switch you from ruminating to thinking.

These examples illustrate that you and everyone else have far more control over brain functions and behavior than is commonly realized. ("I can't help it," is one of the most common phrases in the human language and one of the most erroneous.)

The Prefrontal and Orbital Cortex are where we become the smartest creatures on the planet. It's the seat of our identity, our problem solver, and our primary means of thinking about how to sustain and enhance the best aspects of our existence. Some creatures have larger brains, but they're short on the frontal cortex. As humans we experience, name the experiences, identify, and name the similarities (genus) and differences (differentia) among the experiences to create and think in abstractions. That's our means of understanding and mentally organizing experiences that would otherwise just be a fragmented jumble.

In addition, we have specialized functions of the left-right hemispheres which handle analytic and data synthesis functions respectively. They communicate with one another through a bundle of nerve fibers called the Corpus Callosum. There's still another level of connective nerve fibers surrounding the near surface of the brain that allows data retrieval and transfer throughout all the brain's organs.

As these various organs and divisions of consciousness become better observed and understood, learners and problem solvers become better able to consciously activate the brain lobes most effective for dealing with any topic at hand.

While there are tools for effective thinking, there are also things people do that block or muddle thought.

To think clearly and effectively, the abstract words we use must ultimately refer to perceived concretes. There are many words used that don't refer to anything in existence. They're called "floating abstractions" whose users believe have meaning – usually "higher" meaning which itself is a floating, disconnected abstraction. Popular non-referential abstractions such as "fate," "destiny," "higher good," "racial purity," for example, have no grounding in reality. They are dead ends when used as part of thought processes. In Neoplatonic philosophy, made-up abstractions such as "Being," "Nothingness," "Higher Reality," refer to nothing that exists. They lend themselves to fantasy speculation but do not lend themselves to effective thought.

An often overlooked tool for clear thinking is Aristotelian definition, noting every abstract concept (word) is a member of a class (genus) and also differs from that class (the differentia). An example is the classical definition of humans as animals (the genus or primary group we belong to) with the capacity for abstract reasoning (the differentia, how we differ from other members of that group). Many argue that since most people don't think rationally much of the time, that we're not inherently rational. The fact is that everyone who isn't suffering brain damage has the power to think rationally, once they understand what it is, it's importance, and how to do it.

Most people follow a random associative stream of consciousness much of the time. One thought or reflection leads to another – with no particular purpose beyond whatever is attracting one's interest at the moment. We meander, daydream, ruminate, reminisce and as Nathaniel Branden once said, "Most people don't start thinking until they're in trouble."

On the mid-level of consciousness, we might be working to a purpose and apply data and procedures that experience tells us will get the result we want. This is perceptual level activity such as driving a car or other purposeful but routine activity.

Next higher is the mental process of directed, sequential questioning. We seek, compile, and question the attributes of a situation to acquire information, solve a problem, or achieve a goal. This, by Objectivist definition, is "thinking."

Another mental process is creative visioning, such as contemplating an extremely desirable but currently unreachable result, and imagining what would achieve that result.

This brain-focused epistemology is not commonly taught in school. Although most educators profess to teach students "how to think," you'll be hard pressed to ever see an example in practice. An exception is the "critical thinking" movement in education, which does focus on the questioning process as central to structured thought. (The word "critical" in this context misleads some people. It doesn't refer to being critical such as in negative criticism, it refers to being analytical and focused.)

Here's the capstone to all this, as noted before: **Not only can we choose to question and think or visualize solutions to questions and problems, we can focus and activate the parts of our brain necessary to perform the tasks at hand.** To apply focused attention, you consciously direct your attention to your lower frontal cortex, the Orbital, and you'll start asking questions. (As noted before, turning it on requires no more elaborate activity beyond what it takes to lift a finger.)

To more fully appreciate the importance of the Orbital Cortex, be aware that this is the region where mental health butchers poked ice picks to stir around the brain matter and quiet down inmates in mental hospitals – the infamous "prefrontal lobotomy."

There are plenty of examples of visionary ambitions and analytical processes in this report. Sometimes one person will envision an ideal, but others end up achieving the final result.

For example, Alan Kay envisioned what he called the "DynaBook," a note-book size device that would connect you to all the world's knowledge. That seemed as far out and far-fetched as you could get at the time.

And Ted Nelson envisioned his "Xanadu" which would allow communication and collaboration across all computer platforms. Again, an outrageous vision at that time and thanks to those who followed, particularly Tim Berners-Lee and his World Wide Web, it's a tool many of us use almost every minute of the day.

The concept of "disintermediation" was floated at the MIT Media Lab for many years. Disintermediation meant the elimination of intermediaries who choose what get's published, taught, or otherwise presented to the world. People would be able to connect with one another without editors, publishers, curators, accreditors, and other gatekeepers. And so it has come to pass.

Some other developments evolve indirectly, such as the original vision of PayPal, which, it was hoped, would replace banking. It didn't get there but what it did achieve is hugely successful as a medium for universally convenient financial transactions.

Sometimes the desired vision is as simple as just wanting to streamline a common process, like getting a ride in the fastest, cheapest way possible, or renting a low-cost sleeping room. The vision may be clear and simple, the steps and problems that follow are many and complex. Each step may require multiple levels of collaboration and innovation. Peter Thiel argued from his experience that Ayn Rand's depiction of lone- wolf creators in business and technology was part of the picture, but not complete.

We can expect many divergencies from today's stated intentions. Along with the Kernal company founded by Bryan Johnson, for example, we now also have Neuralink in San Francisco, cofounded by Elon Musk that also aims to develop brain implants for boosting people's mental capabilities. But Brain implants are not likely to be a popularly accepted solution. More likely, less intrusive patches will broadcast to your watch or iPhone to show you regions of your brain to let you observe and increase neural activity. Further on, our most important advances in mental functioning will probably emerge from still unknown and unpredicted technologies.

CHAPTER THIRTEEN   "Real People?"

One criticism of Rand is that her heroes aren't "real people." Real people don't do these things. Actually, she chose real people as templates for characters in her novels.

Howard Roark in *The Fountainhead,* is obviously based on Frank Lloyd Wright. Wright commented that he appreciated her telling it like it is, and he kept *The Fountainhead* for bedside reading.

Her Nathaniel Taggart in *Atlas Shrugged* built his railroad without government subsidy. Subsidies were provided to all other railroads by politicians of the 19th Century. Taggert was modeled after Jim Hill, who built the Great Northwestern Railroad. Hank Rearden, of Rearden Steel, was based on Andrew Carnegie who turned iron into steel. And the "yellow press" newspaper tycoon Gail Wynand in the Fountainhead was based on William Randolph Hearst.

The figurehead hero of *Atlas Shrugged* is John Galt, who invents a motor that runs on "static electricity" providing free electricity to the world. Galt is based on the electrical engineer Nicola Tesla, who gave us alternating current and whose ultimate goal was to tap "ground effect" electricity that would be broadcast like wireless radio providing universal free power to all. (His funders, George Westinghouse and J.P. Morgan, didn't think free electricity for all was a good idea and refused financing.)

Rand's villains are based on people who come and go, but are all too real. These people especially matter when they take on authority and power of government and are then in a position to use the law to trade favors with cronies and exploit or repress the most productive members of society.

Objectivists often perceive themselves as above the norm, and many are, as we see throughout this report. But the most enlightened also acknowledge that like "real" people, they too make mistakes. They strive to learn from errors, do better, and move on.

CHAPTER FOURTEEN   The Future Isn't What It Used To Be . . . It's More

If you want to predict the future, you must invent the future.

Alan Kay

Some, if not most of the Massive Transformative Purposes, will be fulfilled in the years ahead, and each one will be a historic game-changer.

Life-extension, a large part of Silicon Valley sponsored research, for example, would revolutionize medical practice. But research may be slowed or blocked by legislation in response to accusations of the super wealthy gaining longevity at the expense of the masses. In that case, a mix of successful and fake therapies would be available overseas, thus making longevity only for the wealthy a reality.

Space mining could generate staggering amounts of wealth, or succumb to international treaties that might prohibit such development.

Education reform through private universities that threaten existing institutions and practices could be outlawed, as has already occurred in Minnesota, that has outlawed online study with Coursera.

A massive financial recession or depression triggered by federal policies could reduce today's stock market values by up to half. Much of the fortunes of the tech entrepreneurs and investors could disappear overnight.

U.S. military excursions already under way in a dozen countries could escalate to accidental or deliberate war. This could create another level of social and economic disruption and a new wave of "keep your mouth shut" style patriotism.

Commentators who have a limited understanding of technology and business are arguing to have the big tech companies such as Google, Apple, Facebook, Amazon and the like declared monopolies subject to antitrust laws.

If they succeed, the extraordinary investment power of these companies will be dispersed among smaller corporations. Those will be governed by Boards of Directors implicitly expected to keep the companies under control and conventional.

Fewer funds for investment and conservative administrations would bring an end to the Massive Transformative Purpose enterprises such as space colonization, alternative education, and brain and anti-aging research. Business would settle down to the tried and true like the good old days of non-disruption.

In extreme circumstances, specified research and new technologies could be outlawed outright, as has happened repeatedly in U.S. history.

The great, the good, and the not-so-good are all equally possible outcomes in the years ahead.

To be sure, critics of the titans of Silicon Valley are gathering steam.

Accusations are rife with abuses of customer privacy, indifference to Russian troll propaganda channeled throughout the internet, (to Facebook customers in particular), and cooperation with fronts for dictatorships such as China seeking to identify dissidents to imprison.

Objectivist- and Libertarian-led enterprises are not among the offenders and may avoid the worst of the political fallout. But they may nevertheless suffer collateral damage as the public rebels against the tech giants.

CHAPTER FIFTEEN   Oh, Oh, The Dark Side

When you get into bed with government, you're likely to get a lot more than a good night's sleep.

President Ronald Reagan

San Jose is the third-largest city in California, and the southern turf of Silicon Valley. Its primary newspaper, the *San Jose Mercury News,* is no small home-town newspaper, jam-packed as it is with technical/entrepreneurial news.

When its ace journalist, Gary Webb, wrote an expose of the CIA's role in smuggling cocaine from Columbia to support the Contras in the 1990's, the Silicon Valley locals paid attention.

Unlike other segments of our society who take dirty business for granted (and try to grab a share), those with a Libertarian streak, such as many in the Valley, don't buy it.

There are now hints of collaborations among people in Google, Yahoo, Facebook, etc., with U.S. intelligence agencies. If that's true, you can be sure the stories will be told. When they are, the whistleblowers will endure charges of treason, threatening national security, and putting our "brave boys overseas" in danger. If you work in, or collaborate with government, and do unethical or illegal things, that's how you respond to exposure. Ayn Rand made a clear differentiation between business and dirty business, and some in the Valley may have slipped into the dark side.

Other hints don't bode so well either, such as Google, Facebook, Microsoft, and Yahoo yielding to Chinese censorship. They've teamed up with Chinese companies and let them to the dirty work of censorship and identifying and imprisoning dissenters who seek or express, pro-freedom sentiments on the internet. Our major tech companies are facing backlash as misdeeds become publicized by human rights advocates.

One who stands out as honorable at this point is Jimmy Wales, who refused to bow down to China. Wikipedia remains intact, without censorship, for those able to access it. Its 80,000 volunteers will battle to keep it clean.

It's an endless battle. For many years, members of U.S. media who we rely on for information, have participated in government misbehavior.

After the Daniel Ellsberg Pentagon Papers case that exposed the fraud of the Vietnam War, the CIA got smart. They learned they can't shut all reporters up, but they can buy some of them with favors of "special access." Then those media cohorts can help discredit those who reveal criminal behavior by government employees which is what they accomplished in the case of Gary Webb's expose.

"Reliable sources," tell us that today there are journalists at the *New York Times, Washington Post, Chicago Tribune,* and *LA Times,* who collaborate with the CIA in return for exclusive stories from the agency.

This goes back now for many years. In 1977, the *New York Times* identified more than 30 CIA collaborators who worked at 22 major media outlets, including CBS, *Time,* and even the *New York Times*. That was a generation ago, but it's reasonable to assume that U.S. intelligence agencies such as the CIA and NSA have expanded their journalist outreach considerably.

This dirty business is reported at times, but quickly get's shot down. Those who expose it face federal prosecution, and not many are willing to face the beast. Or worse, such as the unfortunate journalist Gary Webb, who amazingly managed to commit suicide by shooting himself two times in the head. "A rare instance," said the coroner. To be sure. Like similar rare cases where suicide was committed by shooting oneself in the back.

Moving to the present day, we are dependent on Ayn Rand-influenced whistleblowers such as Edward Snowden and Chelsea Manning, who forwarded files showing government misdeeds to Julian Assange's WikiLeaks.

After demeaning and brutal treatment in prison and the prospect of a 35-year sentence (commuted by President Obama), Chelsea Manning is stronger now, stronger than her enemies. And she's not finished.

Manning says: "I've read Mises, Hayek, Friedman, and even all of Ayn Rand's books...."

That will help. These will all help Manning maintain courage in the battles ahead.

Similarly, Edward Snowden, an exile in Russia due to his exposure of the misdeeds of the National Security Agency, also gained courage and motivation from his readings of Ayn Rand. An interview with Snowden by Objectivist Jimmy Wales is follows in Chapter Seventeen.

The author of an article in the *New Yorker* declared a deep level of disgust for Snowden, made worse by knowledge that Ayn Rand was part of what the author called "warps and twists" in Snowden's character. (We yet don't know which authors, such as this one, are in bed with the CIA these days. That information awaits the next whistleblower.)

Meanwhile, Julian Assange remains holed up in the Ecuador embassy in London to avoid extradition by the U.S. Government. He declared himself to be Libertarian in an interview with *Reason* magazine. He also hung out from time to time online with Transhumanists and other forward-thinking Libertarian techno freaks. His subsequent behavior strongly reflects these free-thinking influences. If the U.S. government captures him, he'll pay a price.

Some Objectivists with a conservative ethical bent felt it was dishonest to steal information, even information about information being stolen from them. This is similar to Ayn Rand's response to the Ellsberg affair. She avoided commenting on Ellsberg, seemingly not wishing to appear to condone his theft of documents. But she blamed the media for not knowing and reporting about our government's Viet Nam disgrace. She apparently was unaware of the extent of government penetration of the news media at that time.

We have Objectivist and Libertarian influences in the work of Chelsea Manning, Edward Snowden, and Julian Assange, but what does this have to do with Silicon Valley?

Mainly this: Silicon Valley Libertarian/Objectivist-oriented organizations and individuals are uniting to discredit the worst of the Silicon Valley companies. Those who collaborate with dictators or U.S. intelligence agencies, especially those who allow Russian trolls to penetrate and influence U.S. politics, are in for trouble.

CHAPTER SIXTEEN   A Case Study of Objectivist Thinking and Its Results

Imagine a world in which every single person on the planet is given free access to the sum of all human knowledge. That's what we're doing.

Jimmy Wales, Wikipedia

The above statement certainly qualifies as a Massive Transformative Purpose as defined in Chapter Three. And it's the basis of the creation and nonstop growth of Wikipedia.

How does one pursue the ambition of compiling the sum of all human knowledge?

It's not going to be done alone. It'll require participation by researchers on every aspect of human knowledge. The information gathered would have to be evidence-based, subject to scrutiny, and challenge by others. Software and media would have to be found to communicate and continue fact checking. People would have to be willing to participate and contribute content with no financial reward. And since any project of this scope would attract detractors, it would have to prepare to answer criticism, founded and unfounded.

There are precedents, of course. It's immediately obvious that this is to be an updated version of the traditional encyclopedia. Encyclopedias have been an enduring part of Western culture since ancient Greece and Rome. Wales spent his childhood pouring through the pages of an encyclopedia his mother, who could ill-afford it, bought for him.

Another precedent – scientific method – the system of observation checks and balances was invented in the early days of organized scientific endeavor. Scientific method has evolved and every hypothesis and theory is subject to never-ending examination and challenge. When knowledge is generated from observation and experimentation, it's always open to scrutiny and revision as new evidence emerges. The principle is well understood and has been the foundation for virtually all human progress of the past 200 years.

There is also the free minds principle. Progress in the creation of new ideas and knowledge does not occur unless people are free to question and test the status quo. People throughout the world seek knowledge and participate in the creation and propagation of knowledge. But most do so when and where they will not be jailed for their efforts. They do so because curiosity is our unique human motivator. From early childhood, our brains drive us to seek knowledge while those with other agenda, seek to block it.

The software tool for open communication and testing of knowledge is the "Wiki."

Ward Cunningham, considered to be the inventor of the Wiki concept, created it as an open source software tool for sharing knowledge and interests in any realm. Anyone can contribute to a Wiki and anyone can edit what's been contributed.

Jimmy Wales' earliest education was open-ended, basically learn what you want to learn when you want to learn it. He attended a Randolph school, one of those free-mind schools as described in Chapter Eight. He studied finance and epistemology in college, and came upon the writings of Hayek on spontaneous order and especially, the Objectivism of Ayn Rand.

"I'm still very much an objectivist to the core. I think that a lot of the tension people imagine really comes from their not having a deep understanding of some of these ideas."

The Objectivist premise is that by using the tools of reason, logic, and openminded creativity, we can understand every aspect of reality and solve any problem put before us. Compiling and providing all the world's knowledge at no cost to the users is a useful thing to do as the foundation for creating new knowledge. Wikipedia was conceived as a means of putting Objectivist philosophy into practice.

One can look at much of contemporary philosophy for the contrary view, one that disvalues and hence inhibits creative, productive endeavor.

Those who follow Deconstructionism, for example, offer a paradox: anything that one thinks or says about anything is automatically in error, because you can't know everything about something at once. That would mean the premise of Deconstructionism is also wrong. But paradoxes are accepted among many intellectuals as something we have to accept because "real" reality is unknowable.

There are those who question whether there is such a thing as reality or if there is, if it's possible to perceive it. And if there is a reality, there's a question as to whether we or our minds exist or are able to perceive it. These widely accepted paradoxes explain why professional philosophers have added so little to the world's store of knowledge or to our ability to think clearly.

This doesn't make sense to most people and if they get a dose during college, they usually shrug it off as irrelevant to real life. But it has its impact throughout the culture, as among politicians who consider truth to be a convenient fiction and have no problem contradicting themselves from day to the next. They contribute nothing to knowledge, human advancement, or a free society and their followers care not.

There are those who never learned purposeful systematic thinking and urge that we trust our intuition and "think with our heart." In their view, human behavior is inherently emotional and irrational at the root.

Emotions, which seem to overrule reason in many people's lives, are the midbrain's hormonal response to events or ideas that are "for me or against me." They are value based, in other words. Those who hold irrational, contradictory values live in chronic anxiety and perpetual self-contradictions in their emotions and behavior. Thus the most effective psychotherapy these days is values-based, helping clients examine the premises they live by.

For many, grand ambitions and achievements are low, even negative, values.

They may admire a soccer star, for example, whose work they can understand, but are suspicious of someone who wants to do unconventional and seemingly impossible things.

Thus they feel justified in attacking a creative achiever like Jimmy Wales and question his motives when he says: "Wikipedia is first and foremost an effort to create and distribute a free encyclopedia of the highest possible quality to every single person on the planet in their own language.

"I think that the right way to deal with the world is to seek to find ways to survive and prosper and achieve our values. I think it's right for you to do that, and it's right for me to do that.

"And I think that as we go around doing that, there is a right way to deal with each other: with dignity and respect for each other as reasoning human beings."

This fundamental understanding of Objectivism won't sound terribly radical or dangerous to most people. Nor does it seem unreasonable that we need to be free to seek, gain, and share knowledge to expand our powers to enhance our lives.

But there is opposition. Prisons around the world are filled with people who have tried to exert the most basic of human rights, the right to think for oneself.

The freedom and privacy needed to exert human rights are considered expendable among many, even among those at high levels of the U.S. government. Since government efforts to restrict human rights are mostly labeled as official secrets, those who expose them face criminal charges. This battle will never end and Objectivists and Liberians will continue to lead the way.

CHAPTER SEVENTEEN   An Objectivist Interviews a Whistle Blower

Free choice is one of the highest of all the mental processes.
                                                                Maria Montessori

Human history is an endless back-and-forth struggle for liberation.

Free societies create knowledge and wealth. Barbarians swoop in to loot the wealth and destroy the knowledge. Monarchs, priests, and dictators take control. People resist and rebel. The struggles between those who want to live their lives and make their own choices, versus those who want to rule are still played out around the world every day.

Politics are the battleground for people competing for control or liberation. The political movement of Libertarianism emerged directly from an understanding of what repressive laws and traditions mean in terms of shackling human intelligence and progress. Democrats, Republicans, Socialists, and Fascists all pretend to know what's best for the rest of us. Their only redeeming value, in the Libertarian perspective, is the degree to which they keep each other in check.

Objectivists and Libertarians lead the resistance and over the years have taken many risks in the battle for human rights. At present, the greatest government imposition of rights is surveillance which, sadly, is facilitated by advances in information technology.

This author visited South Africa years ago during Apartheid and learned from that nation's leading Libertarian scholars and activists Leon Louw and Francis Kendall about the essence of a police state. It's not a matter of police on every corner, instead, people police themselves. It's people not knowing who the police are, how much they know about you, and who's listening that maintain the repression. People police themselves and their friends and family accordingly.

Unrestricted government surveillance of private communication lays the groundwork for facilitating future intrusions on human rights and freedom. It's hard to find clearer statements of what this is all about than the words of National Security Agency whistleblower, Edward Snowden, as elicited by Jimmy Wales in a January 5, 2018 WikiTribune interview that follows on the next page..

This is an edited version of their conversation over Skype between London and Moscow where Snowden has been since his exposé – unable to return to the United States for fear of prosecution. The discussion ranged across mass spying, journalism, leaking and the risks to privacy from online platforms like Facebook – let alone security agencies vacuuming up all our online communications.

Jimmy Wales: Let's begin by asking where you stand on organizations like Wikileaks releasing huge caches of unfiltered information?

Edward Snowden: I don't pass judgment on whether Wikileaks did the right thing or the wrong thing, because I think this kind of experimentation is important. We need to challenge the orthodoxy.

We need to challenge the presumptions that whatever we're doing right now, the status quo, is the best of all possible worlds. This is the best anybody could possibly do. Instead, we test our premises again and again in different ways, so what I did was I saw that inside the United States government, the National Security Agency had started violating the Constitution in a very unprecedented and indiscriminate way.

They were collecting the phone records, the internet records, all this transactional information about people's private activities: the most intensely intimate and private details of their daily life, without any regard to whether or not they were actually criminal, without any regard to whether there was any problem caused for suspecting they were involved in any wrongdoing whatsoever. Instead, they had developed this new model.

They call it the "collect it all" model where they collect everything they can about every innocent person, so that they have a bucket of everybody in the world's private lives that they can then later sort through and search through at their leisure, if you ever do come to their interest. If you become interesting, they have a kind of surveillance time machine, where they can wind it back, depending on the type of content, the size of it, anywhere from three days to about five years.

JW: What do you think this kind of data-collecting does to the relationship between citizens and the government?

ES: The bottom line here was that the American public was misled in a way that actually matters, because when we think about democracy, when we think about our system of government, every democracy is founded upon a single principle and that's the legitimacy of its government is derived from the consent of the governed.

We cast our votes to make clear our policy preferences, to steer the future of government, but if we're being lied to and our understanding of how the process of government, how the operations of government are being carried out, when our understanding of what the government does in our name, and against us, is not correct, and we vote based on these premises, what's happening is we're starting to lose our seat at the table of government. We're becoming less citizens and more subjects.

JW: What are your concerns over the kind of surveillance in which the U.S. is actively engaged?

ES: In the need to develop a capacity to know what potential enemies are doing, the United States government has perfected a technological capability that enables us to monitor the messages that go through the air. Now that is necessary and important to the United States as we look abroad at enemies or potential enemies. We must know, at the same time, that capability at any time could be turned around on the American people and no American would have any privacy left, such is the capability to monitor everything, telephone conversations, telegrams, it doesn't matter. There would be no place to hide if this government ever became a tyranny, if a dictator ever took charge in this country. The technological capacity that the intelligence community has given the government could enable it to impose total tyranny and there would be no way to fight back. That's the abyss from which there is no return.

JW: People seem to think that monitoring individuals is a recent development. But history has shown us that the U.S. government has long investigated those individuals it has deemed threats.

ES: Two days after Martin Luther King gave his "I have a dream" speech, the FBI's director of domestic intelligence made the judgment that Martin Luther King Jr. was the greatest national security threat facing the United States. This happened in secret. We didn't find it out until years later in the Church Committee. But this is the natural progression of unchecked, unrestrained intelligence agencies in any country. This is not a uniquely American problem. We've seen it happen in other countries before. So the question is, again, how do we do this? We need to have a system of checks and balances. So the idea of this grand bargain was they would create a secret court that would basically issue warrants for intelligence investigations, like they did for traditional criminal investigations. This would be a specialized court that would understand all the difficulties here. They would all have clearances. There would be no fears of leaks and, for a time, that worked. But the problem is, from the 1970s, when the applications that went to this court were few and they were very serious applications, these rule-breaking agencies started to figure out how to abuse the system of rules and the secret court went along with it.

JW: What, if any, role does Congress play in all of this?

ES: Many of the largest critics of surveillance in Congress now aren't allowed to sit in the intelligence-oversight hearings. We have 535 members of Congress, but only roughly 20 of these folks are allowed to actually be briefed on these kinds of domestic surveillance programs, and this is the central problem. You have an executive, a president who doesn't want to be checked and again, this isn't because they're evil. That's because that's what every president does. They don't want the senators sniffing around what they're doing. They don't want courts telling them what they can and can't do.

We see this quite recently with Donald Trump's Muslim ban, which as soon as the courts got wind of it, and had a role and could say, "we have a role to play here," said "look, we understand what you're trying to do here, but it's a violation of the Constitution and you can't." So presidents want to cut people out. The courts increasingly have gone along, as long as the government's said "it's a secret and you can't prove it's actually happening," sorry, we can't help you. And the Congress has been trained over decades to think the best thing for you to do in your re-election campaign is to not look too closely and to open your pockets. [This has] resulted in a system that allowed more than a decade of operation of an unconstitutional domestic dragnet.

JW: Prominent critics of leaks such as John McCain and Hillary Clinton have argued the disclosure of sensitive information puts people at risk. What do you think?

ES: There has never been a case that the government has identified where somebody's been hurt by one of these disclosures, but it's at least theoretically possible. And this is the central question: how should we balance these hard decisions when, on the one hand, we're talking about the theoretical risks of journalism in a free and open society, and on the other, we're talking about the concrete, quantifiable harms of bad policy, of rights violations that are actually backed by evidence?

When we hear people talking too much about how we should do journalism, and not enough about how we should curtail these harmful programs of policies that have been revealed by journalism, that is an indication that people are arguing perhaps less in good faith and are most interested simply in changing the topic of conversation away from what's being done to your rights behind closed doors, as something they can just go "look, this is a political issue," where there may not be any right answer, but we can argue about this until the end of the time, without actually having to confront the criticisms that are being made about how we have ultimately diminished the meaning of rights in the United States and around the world.

JW: On a more meta level, what do you think is more concerning about the internet and the information we involuntarily disclose about ourselves?

ES: We need ways of protecting ourselves from advertisers, as our communications are simply transmitting the internet and yes, it's critically important, particularly in a moment where governments are increasingly unpredictable, unreliable and less representative of what the public actually needs, when governments are becoming less defenders of the public and unfortunately, more oppressors of the public, we need to start thinking about how we can protect our right to protest, how we can actually go to a protest without worrying about our phone, our cell phone, being shown on the roster of the cellphone towers around the protest sites, that the police can simply go "yup, yup, yup, we now have a perfect record of attendance for everybody who had a cell phone that was turned on there."

JW: What lessons can journalists learn from this?

ES: A lot of people like to think of journalism and whistleblowing as two separate topics, but in reality, they are the same issue. We can't have real journalism without informed, reliable sources being able to tell journalists what they need to know, rather than what they are permitted to know, either by policy or process or presidents, or by law. If the law is being broken, if the public's rights are being violated, journalists have to be able to access that material. But if telling journalists about that is itself a crime, now you start raising real questions about all right, how do you actually ensure that this happens? This means, in our world today, whether we like it or not, doing real journalism increasingly relies upon technology and that trend is increasing over time and that means yes, embracing mechanisms like encryption.

JW: What is the main difference in how private data is gathered these days, compared to the past?

ES: Mass surveillance is what the controversy of the last several years is about and this means suspicionless surveillance, whereas the government calls it "bulk collection" of people's records without regard for whether or not they've done anything wrong. This is not normal. This is not something that's happened historically. This is not something that previous societies did. The traditional means of investigation, that we know works, unlike this mass surveillance, which President Obama had two independent commissioners look into and they both said it doesn't work.

JW: There are a number of investigations currently going on into Russian malfeasance during the US election of 2016. "Fake news" is an especially important topic in online news. Do you think censorship works?

ES: Censorship does not do good. We might want to believe it does, and this gets into the fake news problem, for example, that if we just empower Facebook to decide what we can and can't see and what is good and bad, the problem can be solved. But this is a mistake for a number of reasons.

One, it creates a slippery slope where now we have private corporations deciding what can and cannot be said. But further, let's say there are clear cases, we're talking about things like Jihadist propaganda, we're talking about fascist communities that are promoting ideas that are actively harmful, out in the public. The problem is, if you censor them, you don't actually remove them. You don't stop the idea from being spread. You just force them underground. It is underground where these ideas actually propagate best and most effectively. This idea that we can just stamp out ideas, we know does not work. This is the cause of every revolution in human history.

JW: Much of this debate seems to rest on the preconception that if you have nothing to hide, you shouldn't mind any intrusion into your privacy. What do you think?

ES: Privacy's not about having something to hide, privacy's about something to protect. Privacy is the fountainhead of all other rights. Privacy is where rights are derived from, because privacy is the right to the self. Privacy is the right to a free mind. Privacy is the ability to have something, anything, for yourself, for you. Freedom of speech doesn't mean very much if you can't have your own ideas and to have your own ideas, you have to have a safe space to develop these ideas, to figure out what it is that you actually believe in. Then to test these ideas selectively with people you trust, to determine whether this is actually a good idea or whether it's stupid.

If every idea that you had ever uttered was instantly captured and recorded and followed you around for the rest of your life, you would never outlive even the slightest mistakes that you'd made. Freedom of religion, of belief, is not meaningful, it doesn't really exist, if you only inherit the beliefs that came from your family or the people before you or from the state. You have to actually have a chance to read, to look, to try, to experiment with new ideas, to figure out what this life of ours is really about for you.

JW: In recent years we have also seen journalists around the world targeted for their access to confidential sources and information. What implications does this have for the freedom of the press?

ES: Freedom of the press cannot truly, meaningfully exist, unless journalists can contact their sources in absolute confidence and privacy, to inform their understanding of what is actually going on. This goes from the highest levels to the lowest levels of our society. We're not just talking about freedom from unreasonable searches and seizures, by the way. It doesn't matter if the Government is saying, well, we don't listen to your phone calls, if they're actually collecting records of them in the first place, because that seizure of those records itself is unconstitutional. Even again, our lexicon, the words we use, the phrases, private property means something that belongs to you rather than belongs to society. If we do away with privacy, we're doing away with individuality, we're doing away with the self, we're saying that we don't belong to ourselves, we belong rather to society and this is a fundamentally dangerous thing, because when you get into that mindset, where rights don't matter because I'm not using them at this moment, you misunderstand why we even have rights.

JW: Privacy has also allowed women and minorities to organize political campaigns for their rights. In many instances, this has had to be done outside of the eyes of the state

ES: The disenfranchisement of women, saying that they cannot vote, is not right, no matter what the justification, and every progress, every moment beyond then, always started as a minority idea, a minority opinion, and it was privacy that allowed these people to coordinate, that allowed them to develop these ideas and organize people who supported these ideas and to spread these ideas until they reached that point of critical mass that changed the world and made all of us a little more free, that made our lives more fair, that made our future brighter and without this, without privacy, you have created not just an anti-social world, you have created an un-free world and you have not become a bit more safe because of it. Even in prison, the most secured environments we have, people are still assaulted, people are still abused, people are still killed.

CHAPTER EIGHTEEN   Reasonable Expectations

The Next 5 to 10 Years Can be Amazing or Awful, . . . Probably Both

- Objectivists and other innovative thinkers will vastly increase our access to knowledge and find new ways to improve our mental abilities to best utilize that knowledge.

- Libertarians will lead the way in blocking government corruption, surveillance, prosecution of victimless crimes, and repression of minds and markets.

- Focused rationality and visionary creativity will become normal behavior for millions more people.

- Healthy human life spans will be extended significantly, possibly indefinitely.

- Everyone, everywhere will have access to self-managed, accredited education at little or no cost.

- There will still be barbarians at the gate, and political and natural disasters of every kind, but we'll be better equipped to deal with them.

These are reasonable expectations – just extrapolating the history of the past generation. Some developments will occur sooner than 5 or 10 years, depending on exponential acceleration characteristic of today's technologies.

Meanwhile, see the Appendix short list that follows for recommended introductory resources on Objectivism, Libertarianism, mind-expansion, and creative problem-solving.

APPENDIX

Recommended resources:

These will also lead you to other sources that might best suit your personal interests and direction in life.

1) *The Vision of Ayn Rand: The Basic Principles of Objectivism* by Nathaniel Branden. The text of the original lectures that introduced Ayn Rand's philosophy, the best overall summary there is. Available used or on Amazon Kindle.

2) *Introduction to Objectivist Epistemology* by Ayn Rand. The clearest explanation of how the brain interacts with the world and creates knowledge, and the difference between reality-based philosophy and subjectivism.

3) *The Contested Legacy of Ayn Rand: Truth and Toleration in Objectivism* by David Kelly, founder of the Atlas Society.

4) *The Art of Reasoning: An Introduction to Logic and Critical Thinking* by David Kelley.

5) *The Demon-Haunted World: Science as a Candle in the Dark* by Carl Sagan and Ann Druyan. An excellent introduction to the revolution in thinking of the 18th century, which paved the way for all the technological advances of the past 200 years. As accurately described: "Scientific thinking is critical, not only to the pursuit of truth, but to the very well-being of our democratic institutions."

6) *Libertarianism: A Primer* by David Boaz. A comprehensive overview of the history, theories and practical results of Libertarianism.

7) *Economics in One Lesson* by Henry Hazlitt. Eye-opening and highly influential explanation of the results of political economic policy.

8) *The Road to Serfdom* by F. A. Hayek. An excellent introduction to Austrian economics and the theory of free markets.

9) *The Law* by Frederic Bastiat. In 1850, he showed that behavior that would not be tolerated by any individual is no more tolerable if conducted by a group, a majority, or the state.

10) ***The Executive Brain: Frontal Lobes and the Civilized Mind*** by Elkhonon Goldberg. Destined to be highly influential in the study of the nature of human consciousness.

11) ***The Psychology of Self Esteem*** by Nathaniel Branden. How the authentic experience of self esteem depends on one's self-trust and integrity, not social norms and the acceptance/opinions of others.

12) ***The True Believer*** by Eric Hoffer. An examination of why and how people surrender their minds to the control of others.

13) ***Applied Imagination: Principles and Procedures of Creative Problem Solving*** by Alex Osborn. The classic in the field, and responsible for major improvements in business and engineering thinking worldwide.

14) ***The Art of Innovation*** by Tom Kelley and ***Change by Design*** by Tim Brown. Insights and applications of the "design thinking" techniques widely used in Silicon Valley.

15) ***The Mind Map Book*** (and others) by Tony Buzan. "Mind Mapping" is a graphic-based methodology widely used in Silicon Valley by individuals and in group "design thinking" sessions. It has also caught on among many Libertarians and Objectivists as a convenient tool for organizing thoughts, planning, analyzing complex problem situations, and wide-ranging creative brain storming.

16) ***The Universal Traveler: a Soft-Systems guide to creativity, problem solving, and the process of reaching goals*** by Don Koberg and Jim Bagnall. Published in 1972 and republished in 2003 by two professors at the College of Architecture and Environmental Design, California Polytechnic Institute. Typewritten text and distractingly jumbled illustrations don't prevent this from being a much-prized overview of how creative thinking methodologies are created and applied.

17) ***Creative Confidence*** by Tom and David Kelley. How creative culture works at its best in business and education – the IDEO company and the "d. School" at Stanford University.

######

Made in the USA
Lexington, KY
06 July 2019